T0139279

Modern Manufacturing Technology

Modern Manufacturing Technology

Spotlight on Future

Jitendra Kumar Katiyar
Ranjeet Kumar Sahu

CRC Press
Taylor & Francis Group
Boca Raton London New York

CRC Press is an imprint of the
Taylor & Francis Group, an **informa** business

First edition published 2022
by CRC Press
6000 Broken Sound Parkway NW, Suite 300, Boca Raton, FL 33487-2742

and by CRC Press
2 Park Square, Milton Park, Abingdon, Oxon, OX14 4RN

CRC Press is an imprint of Taylor & Francis Group, LLC

ISBN: 9781032066394 (hbk)
ISBN: 9781032066400 (pbk)
ISBN: 9781003203162 (ebk)

DOI: 10.1201/9781003203162

Typeset in Times
by codeMantra

Contents

Preface   xi
Authors   xiii

1 Introduction to Modern Manufacturing Techniques   **1**
  1.1 Need for MMTs   1
  1.2 Classification of MMTs   2
  1.3 Advanced Casting Techniques   3
     1.3.1 Ceramic Shell Investment Casting   4
       1.3.1.1 Applications   4
     1.3.2 Expandable Pattern Casting   5
       1.3.2.1 Applications   5
     1.3.3 Plaster Mold Casting   5
       1.3.3.1 Applications   6
     1.3.4 Ceramic Mold Casting   6
       1.3.4.1 Applications   7
     1.3.5 Vacuum Mold Casting   7
     1.3.6 Slush Casting   7
     1.3.7 Squeeze Casting   8
       1.3.7.1 Applications   9
     1.3.8 Spin Casting   9
     1.3.9 Rapid Solidification   9
     1.3.10 Single Crystal Casting   10
     1.3.11 Stir Casting   11
     1.3.12 Semisolid Casting   11
       1.3.12.1 Thixo Casting   12
       1.3.12.2 Rheo Casting   13
       1.3.12.3 Thixo Molding   14
       1.3.12.4 Strain-Induced Melt Activation (SIMA)   14
  1.4 Advanced Forming Techniques   14
     1.4.1 Explosive Forming   15
       1.4.1.1 Standoff Method   15
       1.4.1.2 Contact Method   15
     1.4.2 Electro-Hydraulic Forming (EHF) 16
     1.4.3 Electromagnetic Forming (EMF)   16
     1.4.4 Peen Forming   17

		1.4.5	Superplastic Forming	18
		1.4.6	Thixo Forming	19
		1.4.7	Ceramic Forming	19
		1.4.8	Glass Forming	20
		1.4.9	Incremental Sheet Metal Forming	21

1.5   Advanced Joining Techniques — 22
    1.5.1   Magnetic Pulse Welding — 24
    1.5.2   Friction Stir Welding — 25
    1.5.3   Explosive Welding — 26
    1.5.4   Electron Beam Welding — 27
    1.5.5   Laser Beam Welding — 27
    1.5.6   Microwave Welding — 28
1.6   Advanced Composites Manufacturing Techniques — 29
    1.6.1   Fabrication Techniques for Polymer-Based Nanocomposites — 30
    1.6.2   Fabrication Techniques for Metal-Based Nanocomposites — 30
        1.6.2.1   Liquid-State Fabrication of MMC — 30
        1.6.2.2   Solid-State Fabrication of MMC — 34
        1.6.2.3   In Situ Fabrication of MMC — 34
    1.6.3   Fabrication Techniques for Ceramic-Based Nanocomposites — 35
        1.6.3.1   Polymer Infiltration and Pyrolysis (PIP) — 35
        1.6.3.2   Chemical Vapor Infiltration (CVI) — 35
        1.6.3.3   Liquid Silicon Infiltration (LSI) — 35
        1.6.3.4   Direct Melt Oxidation (DIMOX) — 35
        1.6.3.5   Sol-Gel Infiltration — 36
        1.6.3.6   Slurry Infiltration — 36
1.7   Advanced Machining Techniques — 36
    1.7.1   Thermal-Assisted Techniques — 37
    1.7.2   Plasma-Assisted Machining — 37
    1.7.3   Laser-Assisted Machining — 40
    1.7.4   Electron Beam Assisted Machining — 40
    1.7.5   Ion Beam Machining — 40
1.8   Advanced Finishing Techniques — 41
1.9   Nanoparticles Production Techniques — 44
1.10   Future of Advanced Manufacturing — 44
References — 45

2   Additive Manufacturing – "An Evolutionary Pace"   47
2.1   Background of Additive Manufacturing — 47
2.2   Glimpse of Non-metal Additive Manufacturing Techniques — 48

2.2.1   Stereolithography (STL)   50
2.2.2   Selective Laser Sintering (SLS)   50
2.2.3   Laminated Object Manufacturing (LOM)   51
2.2.4   Fused Deposition Modeling (FDM) or Fused
        Filament Fabrication (FFF)   52
2.2.5   Solid Ground Curing (SGC)   52
2.2.6   3D Ink-Jet Printing   53
2.2.7   Multi-Jet Modeling (MJM)   53
2.3   Advent of Metal Additive Manufacturing Techniques   55
2.4   Conceptual Realization of Metal Additive
      Manufacturing Techniques   56
      2.4.1   Powder Bed Fusion Technique   57
      2.4.2   Direct Energy Deposition Technique   58
      2.4.3   Metal Binder Jetting Technique   60
      2.4.4   Bound Powder Extrusion
              (Bound Powder Deposition) Technique   61
      2.4.5   Wire Arc Additive Manufacturing Technique   62
      2.4.6   Promising Applications of Metal 3D Printing   64
2.5   Future Directions   65
2.6   Summary   66
References   66

3 Advanced Machining in the Age of Nanotechnology   69
3.1   Nanotechnology in Current Millennium   69
3.2   Science of Nanoparticles   71
3.3   Glimpse of Nanoparticle Applications   74
3.4   Nanoparticles Production Techniques   75
      3.4.1   Ball Milling   75
      3.4.2   Wire Explosion   77
      3.4.3   Inert Gas Condensation   77
      3.4.4   Sputtering   77
      3.4.5   Submerged Arc Nanoparticle Synthesis
              System (SANSS)   78
      3.4.6   Combustion Flame   78
      3.4.7   Plasma Processing   78
      3.4.8   Spray Pyrolysis   78
      3.4.9   Solution Precipitation   79
      3.4.10 Sol-Gel Processing   79
      3.4.11 Micro-Emulsion   79
      3.4.12 Polyol Technique   80
      3.4.13 Aerosol Synthesis   80
      3.4.14 Chemical Vapor Condensation   80
      3.4.15 Boiling Flask-3-Neck   81

		3.4.16 Microfluidic Reactor	81
3.5	Exploitation of Nanotechnology Concept in Advanced Machining		81
	3.5.1	Pulsed Laser Ablation	82
	3.5.2	Electrochemical (or Electrochemical Micromachining) Technique	83
	3.5.3	Micro-Electrical Discharge Machining (Micro-EDM)	84
	3.5.4	Micro-Electro Chemical Discharge Machining (Micro-ECDM)	86
3.6	Challenges of Nano-Based Manufacturing		89
3.7	Summary		89
References			89

4 Ultrafine Electronic Devices Manufacturing Techniques   **95**

4.1	Semiconductor		95
	4.1.1	Semiconductor Doping	96
		4.1.1.1 Diffusion Technique	96
		4.1.1.2 Ion Implantation	96
4.2	Importance of Semiconductors		97
4.3	Materials		97
	4.3.1	Silicon Wafer Fabrication	97
		4.3.1.1 CZ Crystal Growing	98
		4.3.1.2 Floating Zone	98
4.4	Measurement of Wafer Characteristic		98
	4.4.1	Hot Point Probe	98
	4.4.2	Four-Point Probe	99
	4.4.3	Fourier Transform Infrared Spectroscopy	99
4.5	Oxidation		99
4.6	Deposition		100
	4.6.1	Physical Vapor Deposition (PVD)	100
		4.6.1.1 Evaporative Deposition	101
		4.6.1.2 Electron Beam PVD	101
		4.6.1.3 Sputtering	101
	4.6.2	Chemical Vapor Deposition (CVD)	101
4.7	Lithography		102
	4.7.1	Photolithography	102
4.8	Etching		103
	4.8.1	Wet Etching	103
	4.8.2	Dry Etching	104
		4.8.2.1 Sputtering	104
		4.8.2.2 Plasma Etching	104

4.8.2.3   Reactive Ion Etching   104
4.8.2.4   Cryogenic Dry Etching   105
4.9   Metallization   105
4.10   Wire Bonding   106
4.10.1   Ball or Capillary Bonding   106
4.10.2   Wedge Bonding   106
4.11   Packaging   107
4.11.1   Through Hole Mount Package   107
4.11.2   Surface Mount Package   107
4.11.3   Chip-Scale Package   107
4.11.3.1   Wafer Level Chip-Scale Package   108
4.11.3.2   Wire Bonded Ball Grid Array   108
4.11.3.3   Flip-Chip Ball Grid Array   108
4.12   MEMS Technology   108
4.12.1   Materials for MEMS   109
4.12.2   Fabrication Process   109
4.12.2.1   Bulk Micromachining   110
4.12.2.2   Surface Micromachining   110
4.12.2.3   LIGA   110
4.12.3   Application of MEMS   111
4.12.4   Issues of MEMS   112
4.12.5   Future of MEMS   113
4.13   NEMS Technology   113
4.13.1   Materials and Fabrication of NEMS   114
4.13.2   Application of NEMS   114
4.13.3   Future of NEMS   115
4.14   Summary   115
References   116

5   A New Vista of Manufacturing Technology
    for Industry 4.0   119
5.1   Background and Impact of Industrial Revolutions   119
5.2   Introduction to Industry 4.0   120
5.3   Transforming Landscape in Modern Manufacturing   121
5.4   Major Challenges   125
5.5   Case Study   126
5.6   Summary   126
References   126

Index   127

Preface

Over the past few decades, throughout the global industrial market, there is a continuous struggle for innovative design and manufacturing of high-quality near-net shape products with less cost in minimal time, processing of exotic materials and production of ultrafine materials and products. These stringent requirements put excessive pressure on the capabilities of conventional manufacturing techniques such as casting, forming, joining and traditional machining to achieve the same. Therefore, to meet these challenges, newer manufacturing techniques, known as modern manufacturing techniques (MMTs), have emerged in the industry as efficient and economic alternatives to conventional ones. Studies on the MMTs have become considerably important to realize their applications in various scientific fields and the way to accomplish the need for complexity of product design and manufacturing, surface integrity and ultrafine features. The present book describes the background of manufacturing techniques in brief followed by the advent of and introduction to MMTs. The broad classification of MMTs is concisely elaborated in the book, so as to provide readers a systematic and coherent picture of the various techniques established in recent years that provide many breakthroughs at present and also into the future. Among the various types of MMTs, additive manufacturing (AM) has grown by leaps and bounds in current times owing to significant interest in the manufacturing sector in developed and developing nations and is still in its evolutionary pace. This book provides a conceptual overview of AM technologies, starting with the fundamentals of layered processing of polymers to the product formation and their promising applications, so readers can grasp these concepts quickly. Then, some of the recent exciting developments of AM technologies using metallic and advanced materials and their important application areas are discussed in detail in the book.

It is known that for micro-scale manufacturing, the advanced machining techniques have become inevitable and popular MMTs. These techniques are capable of machining microfeatures on various exotic engineering materials such as superalloys, carbides and ceramics along with the general-purpose engineering materials such as copper, aluminum, stainless steel, etc. to very high accuracy and precision. The microfeatures have found widespread applications in many industrial domains. However, with current growing trends toward features at the nanoscale (ultrafine features), i.e., nanomaterials owing to their exceptional properties and interdisciplinary emerging applications, production of nanomaterials with the desired size, shape and distribution using various physical and chemical techniques has become increasingly important. The role of nanomaterials technology in the present century and the comprehensive overview of the physical and chemical techniques used for nanomaterials production is explained briefly in the book. Moreover, the advent of nanotechnology concepts in advanced machining techniques and the recent exploration of some of these techniques (coming under the umbrella of physical techniques) for production of nanomaterials are addressed in detail. In addition, the importance of conceptual realization of these advanced machining techniques, the methodology of production of nanomaterials and their characterization using certain diagnostic methods are also discussed. Further, fabrication of ultrafine electronic devices including micro-electro mechanical systems (MEMS), nano-electro mechanical systems (NEMS), semiconductors and optical systems has played a considerable role in society due to their enhanced system overall efficiency. The mechanisms of fabrication of the above-mentioned devices using the present sophisticated techniques are explored.

In the 21st century, most of the manufacturing industries require higher automation and high-energy efficiency manufacturing systems that could assure high levels of sustainability and productivity. The movement toward digitization in manufacturing will revolutionize the manufacturing industry by bringing new functionalities that will change the rules of the game for the industry players. Therefore, today, the scientific innovations and technology advancements unfold new vistas of manufacturing engineering where seamless integration of advanced manufacturing systems, automation, novel materials, ultrafine electronic products, supplier, logistics and mass customization into the information network would open up immense opportunities to roll out a plethora of newer manufacturing techniques tomorrow, and these should create the bliss of Industry 4.0. Industry 4.0 could emphasize the idea of consistent digitization and link all productive units in an economy. This book will vividly excite the imagination of the audience about the new vista of manufacturing technology of tomorrow that shall drive Industry 4.0 of today.

Authors

Jitendra Kumar Katiyar, PhD, is a Research Assistant Professor in the Department of Mechanical Engineering, SRM Institute of Science and Technology, Kattankulathur, Chennai, India. His research interests include tribology of carbon materials, composites, self-lubricating polymers, lubrication tribology, modern manufacturing techniques and coatings for advanced technologies. He earned his Bachelor's degree with honors from UPTU Lucknow in 2007. He earned his Master's degree from the Indian Institute of Technology, Kanpur, India in 2010 and his PhD from the same institution in 2017. He is a life member of Tribology Society of India, Malaysian Society of Tribology, Institute of Engineers, India, The Indian Society for Technical Education (ISTE), among others. He has authored/co-authored 30+ articles in reputed journals, 30+ articles in international/national conferences, 12+ book chapters and 5+ books published by reputed publishers such as Springer and CRC Press, USA. He has served as a guest associate editor for special issues in *Tribology Materials, Surfaces and Interfaces*, Part J: *Journal of Engineering Tribology*, *Arabian Journal for Science and Engineering*, *Industrial Lubrication and Tribology* and review editor (Tribology) in *Frontiers in Mechanical Engineering*. Further, he is member editorial board in *Tribology Materials, Surfaces and Interfaces*. He is also an active reviewer in various reputed journals related to materials and tribology. He has delivered more than 30 invited talks on various research fields related to tribology, composite materials, surface engineering and machining.

Ranjeet Kumar Sahu, PhD, is an Assistant Professor in the Department of Mechanical Engineering, National Institute of Technology, Karnataka, Surathkal, India. He earned his BE degree in Mechanical Engineering in 2002 from Berhampur University, MTech degree in Production Engineering in 2011 from NIT Rourkela and PhD in Nanomanufacturing from Indian Institute of Technology, Madras, in 2016. As a credential, he received Institute Day Best Scholar Award in 2010, Best MTech Award in 2012 and Prof. M.S. Shanmugam Best PhD Thesis Award in 2016. He has published many journal papers, books and book chapters at national and international levels. His current research areas of interest include micro/nanomachining, nanomaterials synthesis and characterization, precision engineering and additive manufacturing.

Introduction to Modern Manufacturing Techniques

1

1.1 NEED FOR MMTs

Conventional manufacturing techniques have met the requirements of industry for the past few decades. However, to become successful in today's industry, manufacturing has to meet the increasing demands in the market, in particular, the demands for innovative design and manufacture of high-quality near-net shape products with less cost, increased productivity and reduced wastage of material, exotic materials processing and ultrafine materials and products. But, conventional manufacturing techniques could not meet the above demands. Industrial researchers have made continuous efforts to advance manufacturing technology. So, modern manufacturing techniques eventually emerged as efficient manufacturing techniques to meet the shortcomings of the conventional manufacturing techniques. Therefore, today, the scientific innovations and technology advancements unfold new vistas of manufacturing engineering where seamless integration of advanced manufacturing systems, automation, novel materials, ultrafine electronic products, supplier, logistics and mass customization into information networks would open up immense opportunities to roll out a whole lot of newer manufacturing techniques.

DOI: 10.1201/9781003203162-1

1.2 CLASSIFICATION OF MMTs

Modern manufacturing technologies are classified to meet the various demands fostered in the market, in particular, the demands for pioneering design and manufacture of high-quality near-shape products with minimum cost, increased productivity and reduced wastage of material, exotic materials processing and ultrafine materials and products. The MMTs are classified into six major manufacturing sections such as advanced casting, advanced forming, advanced machining, advanced joining, modern manufacturing of composites, modern manufacturing of semiconductors and micro-electro mechanical systems (MEMS) and additive manufacturing. Further, these sections are divided into subsections. The detailed classification is shown in Figure 1.1.

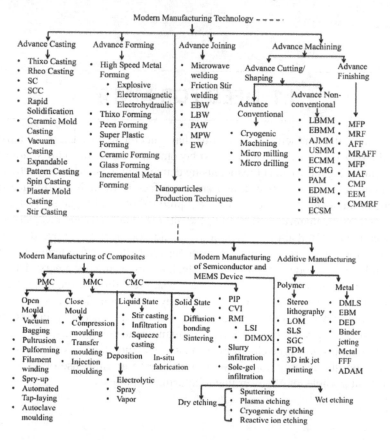

FIGURE 1.1   Classification of modern manufacturing technology.

ADAM, atomic diffusion additive manufacturing; AFF, abrasive flow finishing; AJMM, abrasive jet micromachining; CMC, ceramic matrix composite; CMMRF, chemo-mechanical magneto-rheological finishing; CMP, chemo-mechanical polishing; CVI, chemical vapor infiltration; DED, direct energy deposition; DIMOX, direct melt oxidation; DMLS, direct metal laser sintering; EBM, electron beam melting; EBMM, electron beam micromachining; EBW, electron beam welding; ECMG, electro-chemical micro-grinding; ECMM, electrochemical micromachining; ECSM, electro-chemical spark machining; EDMM, electric discharge micromachining; EEM, elastic emission machining; EW, explosive welding; FDM, fused deposition modeling; IBM, ion beam machining; LBMM, laser beam micromachining; LBW, laser beam welding; LOM, laminated object manufacturing; LSI, liquid silicon infiltration; MAF, magnetic abrasive finishing; MFP, magnetic float finishing; MMC, metal matrix composite; MPW, magnetic pulse welding; MRAFF, magneto-rheological abrasive flow finishing; MRF, magneto-rheological finishing; PAM, plasma-assisted machining; PAW, plasma arc welding; PIP, polymer infiltration and pyrolysis; PMC, polymer matrix composite; RMI, reactive melt infiltration; SC, squeeze casting; SCC, single crystal casting; SGC, solid ground curing; SLM, selective laser melting.; SLS, selective laser sintering; USMM, ultrasonic micromachining.

1.3 ADVANCED CASTING TECHNIQUES

Casting is a very old process. In the 8th century BC, the Chinese first made cast iron. But after the invention of the blast furnace by Europeans in the 14th century, the production of cast iron was increased (Manikanda and Vignesh, 2017). Further, it is interesting to know that China is continuing its position as a leader in casting production worldwide (according to the 49th census of world casting production 2014) (Soiński et al., 2016). It is a manufacturing process in which molten material is poured into a mold that contains with/without hollow cavity of the desired shape and then permitted to solidify. The obtained solidified part by ejection or breakage of mold is known as casting. It is used to fabricate the complex shape parts that are not possible by other manufacturing methods. Therefore, it is very important to improve the production and productivity of industry to meet the various demands of fostering in the market. Hence, the advanced casting methods are introduced by industries over conventional casting methods (sand casting, shell mold casting, investment casting, die casting, pressure die casting, centrifugal casting). The advanced casting methods are shown in Figure 1.1. A brief discussion of each advanced casting method is given in the following sections.

1.3.1 Ceramic Shell Investment Casting

In investment casting, the wax pattern is dipped in a refractory aggregate before de-waxing but in ceramic shell investment casting, a ceramic shell is developed around a tree assembly through frequent immersion a pattern into a slurry, i.e., refractory material like zircon with binder. After each immersion and stuccoing is accomplished, the tree is allowed to thoroughly dry before applying the next coating. Hence, a shell is developed around the tree. The thickness of this shell is dependent on the size of the castings and temperature of the metal to be poured. After the development of ceramic shell, the entire assembly is employed into an autoclave or flash fire furnace at a high temperature. The shell is heated to ~900°C to burn out any residual wax followed by the development of a high-temperature bond in the shell. Further, it can be stored for future use or molten metal can be poured into them immediately. If the shell molds are stored, then preheating is required before molten metal is poured into them (Prasad, 2012).

Advantages
1. Draft allowance and rapping is not required because it is not taken out from the mold.
2. Require less time. Further, core, parting line and riser are not required.
3. Molding flasks are inexpensive, i.e., polystyrene.
4. Minimal cleaning and finishing is required.

Disadvantages
1. It is not suitable for machine molding.
2. The product is widely used for product research in the foundry industry due to destruction of pattern.

1.3.1.1 Applications

This process is widely used in the fabrication of cylinder heads, crankshafts, brake components and machine bases. Further, Polymethylmethacrylate (PMMA) and polyalkaline carbonate are used as pattern materials for ferrous casting.

1.3.2 Expandable Pattern Casting

This method is also known as the lost foam process, lost pattern process, evaporative foam process, or full mold process. In this process, polystyrene foam pattern, closely packed by sand is used that is vaporized by application of molten material into it (Monroe et al. 2008). The foam pattern consists of the sprue, risers, gating system and internal cores (if required). More importantly, it does not require cope and drag sections. This process is comprised of three basic steps. In the first step, coating of refractory material is carried out over polystyrene foam pattern using the spraying process. In the second step, coated foam pattern is placed in mold box and closely packed by sand. The molten material is poured into the portion of pattern in the third step. The evaporation of the foam pattern is started as it comes into contact with molten material because this the mold cavity is filled.

Advantages
1. Pattern is not needed to remove from mold.
2. Cope and drag are not required making it very simple and very fast fabrication of mold.

Disadvantages
1. Every time new material for the pattern is required.
2. The cost of production is widely dependent on the cost of producing foam pattern.

1.3.2.1 Applications

This process is widely used in the mass production of casting for automobile engines.

1.3.3 Plaster Mold Casting

It is also known as the metalworking casting process that is similar to sand casting except molding materials (Charles, 1992). Here, the Plaster of Paris (POP) is used as a molding material. To improve the green strength, dry strength, permeability and castability, the additives are added in POP. Importantly, it is used only to cast the non-ferrous metals. The process is very simple in which POP is homogeneously blended and sprayed over the pattern followed by shaking that helps to fill the small features by POP.

Further, it is kept for 15–20 minutes to dry and then the pattern is removed. The developed mold is then backed between 120°C and 250°C to remove the moisture followed by the assembly of dried mold and preheating. On preheated mold, molten material is poured and kept for solidification. Finally, mold is broken to remove the cast part.

Advantages
1. Minimal scrap material is produced.
2. Very economical for complex shape parts.
3. The minimal cross-section of up to 0.6 mm can be achieved.
4. Small casting as 30 g as well as large casting as 45 kg can be produced.

Disadvantages
1. This process is only used for lower melting point non-ferrous materials such as Al, Mg, Cu and Zn.
2. Production volume is restricted due to long duration in solidification.
3. Close monitoring of parameters of POP such as plaster composition, pouring procedure and curing techniques is required.
4. It is not as stable as sand.

1.3.3.1 Applications

This process can be used to fabricate small gears, fittings, tooling, lock components, valves, ornaments, aircraft parts, small propellers, handles, small housings, etc. Further, it can be used to develop the prototype and short run production of Al and Zn components.

1.3.4 Ceramic Mold Casting

This process is used to cast the components at a much higher temperature. Refractory sand is used for making mold (Groover, 2018). The refractory sand is homogeneously blended silica grain with ethyl silicate, water, alcohol and a gelling agent such as HCl. The prepared slurry is then poured around the pattern and allowed to solidify for 5–6 minutes. Further, the pattern is removed and fired by using alcohol to improve strength and rigidity. The firing of mold has also created the network of microcracks. These cracks improve the permeability and collapsibility properties of ceramic mold. After the development of two halves, it is assembled and preheated before pouring the molten material. The final cast part is removed by breaking the ceramic mold.

Advantages
1. It can produce parts with thin sections.
2. It can produce a part with excellent surface finish and higher dimensional accuracy that eliminates machining.
3. It can be used to fabricate parts made of ferrous materials and higher melting point metals.
4. Pattern is reusable and cheap. Further, there is no restriction on the casting size.

Disadvantages
1. It is a relatively expensive process.
2. Mold preparation is time-consuming.

1.3.4.1 Applications

This process is used to cast the parts such as impellers, complex cutting tools, plastic mold and tooling.

1.3.5 Vacuum Mold Casting

This process was developed by Japan in 1970 to improve production. The term "vacuum" refers to mold making rather than casting operation itself. In this process, chemical binders are not used in sand but the sand mold is held together by vacuum pressure (Groover, 2018).

Advantages
1. Due to no binder in sand, the easy recovery of sand is possible. Further, sand is not required for mechanical reconditioning.
2. Moisture defects are eliminated because no water is used in process.

Disadvantages
1. It is a time-consuming process because the whole process is carried out by vacuum.
2. It is not widely adaptable to mechanization.

1.3.6 Slush Casting

Slush casting is a variation of the permanent mold process. In this process, metal is allowed to remain in the mold until the shell of the desired size is formed. Then, mold is inverted, and the remaining metal is poured out.

Finally, the mold halves are separated to eject the hollow casting of an excellent surface finish. It is important to note that this process can achieve a variable wall thickness. This process is used to cast low-melting-temperature metals into ornamental objects such as candlesticks, lamp bases and statuary (Groover, 2018).

1.3.7 Squeeze Casting

This process is also known as liquid metal forging (Groover, 2018). It is a combination of the casting and forging process. In this process, the molten material is poured into open face die. Then, punch is advanced toward the die and metal by applied pressure that is less than the forging pressure followed by solidification. After that, punch is retracted and the cast part is ejected out with the help of an ejector pin. Furthermore, to operate the casting process efficiently, there are various parameters that need the operator's attention. These are casting temperatures that depend on alloy and cast geometry (the starting point is generally 6°C–55°C above the liquidus temperature), tooling temperature (~190°C–315°C), pressure level that is kept in between 50 and 140 MPa and finally lubricant (most widely used is colloidal graphite lubricant in casting of Al, Mg and Cu alloys) that is sprayed on the warmer dies before casting. The developed cast product has shown the highest mechanical properties. It is a very simple, economical and efficient casting process in its use of raw materials.

Advantages
1. Very minimal or no machining is required on the cast part.
2. Porosity of the cast part is very low. Further, a good surface texture is obtained.
3. There is 100% utilization of materials so no wastage.
4. Fine microstructure with higher strength in components can be obtained.

Disadvantages
1. Complex tooling is used that increases the cost. Further, tools are dedicated to specific components only.
2. Very precise control of pressure is required.
3. It can be used when high production volume in the industry is required.

1.3.7.1 Applications

This process is used to develop the Al automotive wheels and pistons, and gear blanks made of brass and bronze. Further, it has commercially prospered in making manufacturing parts including an Al-dome, ductile iron mortar shell, stainless steel blades, superalloy discs and a steel bevel gear.

1.3.8 Spin Casting

It is an advanced process of centrifugal casting (Hashmi et al., 2014). This process used rubber as mold material due to which it is also known as centrifugal rubber mold casting (CRMC). Then, the metal is poured into the rubber mold that is spin around the central axis followed by solidification of metal. It is a widely used non-expandable process that has shown a good surface finish. Further, it is an ideal process for quick and economical production of fully functional fragile metal or plastic products.

Advantages
1. Fast turnaround time makes it more competitive.
2. Tooling cost is lower. Further, it is very simple to use technology.
3. It does not require a heavy mold cover plate to handle each cycle.

Disadvantages
1. Here, zinc die-casting alloy as material can be used due to its lower melting point. Further, it has higher strength and hardness.
2. Rubber mold limits the size and complexity of products.

1.3.9 Rapid Solidification

This process is used to fabricate amorphous alloys. Rapid solidification involves a higher cooling rate of fluid (~10^6 K/s) due to which the molten material does not get sufficient time to crystallize along the long distance (Hashmi et al., 2014; Groover, 2018). Further, the crystal is grown at a much faster rate (250 m/s) due to which it is attained the level of nucleation rate. Thus, the liquid is solidified without developing the crystalline grains. Such type of alloy is called amorphous alloy or metallic glasses. Moreover, it is noticed that the significant extension of solid solubility, grain refinement and minimal micro-segregation during rapid solidification. The amorphous alloy consists of Fe, Cu and Cr that are alloyed with C, P B, Al and Si. They exhibited improved corrosion resistance, good ductility and higher strength. Furthermore, they exhibited high permeability, high resistance to eddy current and minimal loss of magnetic hysteresis. Therefore, this process is widely used to develop magnetic steel cores for transformers, generators, motors, lamp ballasts, magnetic amplifiers and linear accelerators. Additionally, its major application is to consolidate the near-net shape structure of aerospace engines by using rapidly solidified powders. These alloys are developed in the form of powder, wire, ribbon, strip and fiber.

This process is further divided into two categories. These are roll casting and melt spinning. The roll casting is the most popular rapid solidification process to develop advanced materials. Either single roller or pair of chilled rollers can solidify the materials. Further, the single roller process is also known as melt spinning, and if two rollers are used in the process, then it is known as twin-roll or strip casting. Although both the processes have physical similarities, melt spinning is generally used in the laboratory scale for research, while twin-roll is generally used in the industry.

In melt spinning, the alloy is melted in a ceramic crucible through induction coils that is propelled at a very high speed under high gas pressure against a rotation of copper chill blocks. These chills solidify the melt rapidly and form the amorphous alloy. Further, the twin belt caster process includes melt feeding, belt stabilization and control, heat transfer control and mold tapering. The applications of this process include sheets, strips and tubes.

1.3.10 Single Crystal Casting

In this process, the mold is prepared that is having one heated end and another chilled end followed by the longest cooling distance due to which the metal is cast. The solidification is started at chill plates, and due to slow pulling of the metal, the dendrites are grown as a part of long dendritic crystals toward the heated end (Hashmi et al., 2014; Groover, 2018). There are two methods such as crystal pulling method and floating zone method that are used to develop the single crystal.

The crystal pulling method is also known as Czochralski crystal growth process. In this process, a seed crystal of a required diameter is dipped into the molten material and then slowly pulled at a rate of 10–20 μm/s with a small rotation of 1 rev/s. Pulled material is started to solidify on seed crystal and the crystal structure of the seed is continued throughout. This process grows the single crystal of Si and Ge. The ingot of a single crystal is obtained typically 1 m length and 50–150 mm diameter. It is widely used to develop the single crystal gas turbine blade of Ni-based superalloys. The other methods to develop the turbine blades are conventional casting and directional solidification process. Furthermore, the floating zone method is widely used to develop the single crystal for the fabrication of microelectronics devices. In this process, the polycrystalline silicon is kept on a single crystal and then moves slowly in the upward direction to pass through the induction coils, due to which single crystal grows in the upward direction while maintaining its orientation. Finally, a thin wafer is cut from the rod followed by cleaning and polishing. There are a wide range of applications of single crystal components including filters, windows and sensors for security systems and satellite equipment, antennas, laser components, scanners for brain and heart examination, telecommunication filters, resonators and oscillators, and optical components.

1.3.11 Stir Casting

Stir casting, a liquid-state method, is used to fabricate the metal matrix composite (MMC). For the development of MMC, the disperse phase of fiber or particles is blended in the molten metal matrix by use of stirring process. The stirring of molten metal is performed continuously to eliminate the air bubbles and agglomeration of disperse phase. Then, it is allowed to produce the MMC by means of either conventional casting or conventional metal forming process (Sahu and Sahu, 2018).

Advantages
1. It is a very simple and most economical process.
2. Better mixing of disperse phase is obtained due to the high viscosity of semisolid liquid.

Disadvantages
1. The cluster formation inside the liquid is observed at a higher concentration of the dispersed phase.
2. Due to the density difference in dispersed phase and matrix, it is observed a gravity segregation of dispersed phases.

1.3.12 Semisolid Casting

It is a relatively newer technology in the field of casting. It offers various advantages over the near-net shape manufacturing method. In this process, the cast part is produced by using the semisolid slurry that is kept between solidus and liquidus temperature (Kirkwood et al., 2010). It is ideally suitable for die casting and allows for near-net shape forming that reduces machining. This process is the combination of liquid metal casting and solid metal forging. It is divided into four major categories. These are thixo casting, rheo casting, thixo molding and strain-induced melt activation (SIMA).

1.3.12.1 Thixo Casting

The prefix "Thixo" is derived from thixotropic that means the viscosity is decreased with respect to time (Viswanathan et al., 2008). Further, it is a general term used to depict the near-net shape forming process from semisolid non-dendritic melts. The electromagnetic stirring is used to cast the non-dendritic billets by producing a minimum shear rate. It is widely dependent on stirring intensity, excitation frequency and cooling rate. Further, a pre-cast billet that is having a non-dendritic structure is heated up to semisolid stage using an induction furnace and then insert into the mold cavity with the help of die-casting equipment. Finally, the cast parts are obtained by applying pressure on billets in die casting. This method is widely used to cast the parts of non-ferrous materials (Groover, 2018).

Advantages
1. It produces complex shape structures using only one step forming due to low yield strength, high fluidity and low forming load.
2. It reduces shrinkage allowing achievement of closer dimensions.
3. It is an energy-efficient process. Further, the production rate is similar to pressure die casting.
4. The achieved surface quality is suitable for plating.

Disadvantages
1. The operator requires a higher level of training and skill.
2. The process is widely dependent on temperature. It needs to be optimized for an efficient process.

1.3.12.1.1 Applications
Thixo casting is widely used to develop aluminum parts for various applications such as disc brake calipers, engine pistons, antilock brake valves, automotive wheels, engine suspension mounts, air manifold sensor harnesses, engine blocks and oil pump filter housings.

1.3.12.2 Rheo Casting

The prefix "Rheo" is derived from the rheology, it means the science of deformation and flow of material. When the semisolid material is stirred to obtain the composite, this process is known as rheo casting (Viswanathan et al., 2008). Here, the slurry is cooled up to the semisolid state and agitated to prevent dendrite formation, and then it is introduced in high-pressure die without any intermediate solidification steps. Due to which, it is allowed to cool the semisolid materials into the desired parts. This process is similar to conventional die casting. This process is also used to fabricate aluminum parts, and it consists of four basic steps. These are stirring, dendrite fragmentation, pressure waves and numerous solidification nuclei.

a. **Stirring:** The liquid aluminum is stirred and allowed to cool up to the semisolid stage.

b. **Dendrite fragmentation:** It can be achieved by the variation of stirring process and allowed to cool below liquidus temperature that produces several small solid fragments with globular-shaped aluminum particles.

c. **Pressure wave:** It is generated in the runner to penetrate the semisolid structure.

d. **Numerous solidification nuclei:** Here, the liquid is poured into the container at above the liquidus temperature and allowed for rapid cooling that permits the formation of a large number of solid nuclei. It prevents the formation of dendrite instead of a large number of globular solid particles.

Advantages

1. Low pressure is used in comparison to die casting. Further, it can be produced equal or more parts from die casting. It can be produced equal or more parts in comparison to die casting.
2. It can be easily automated and consistent.
3. It produces a uniform microstructure with no air entrapment and shrinkage.

Disadvantages

1. It requires a high level of technology and a skilled operator.
2. It requires a relatively higher feedstock cost.
3. It requires precise control of operating conditions.

1.3.12.3 Thixo Molding

Due to the increased demand for electronic and communication parts such as laptop computers, cameras, projectors and cell phones in the late 1990s, thixo molding was developed by Japan. Further, it was widely adopted by Taiwan and now by China. Further, this process is used to cast magnesium (Mg) metal (Viswanathan et al., 2008). The process is similar to injection molding. The granules of Mg alloys are fed into barrel that is propelled by a rotating screw. These screws are heated at a semisolid temperature range and allowed to blend the granules properly followed by injection into mold cavity.

1.3.12.4 Strain-Induced Melt Activation (SIMA)

If the processing parameters are not properly controlled in rheo casting, then porosity and segregation issues arise in the casted part that is extensively reduced by using the SIMA process because it is a combination of casting and rolling processes. In this process, the required alloy morphology is obtained through deformation followed by heat treatment. It is a promising technology due to its several merits such as easy and minimal setup cost. Further, it can be used to cast metal alloys including Al, Cu, Mg and ferrous alloys (Viswanathan et al., 2008).

1.4   ADVANCED FORMING TECHNIQUES

The conventional forming techniques such as rolling, extrusion, forging and drawing are well known where the large plastic deformation is occurred either by manual or mechanical means to get the desired shape and geometry. The desired geometry has obtained by hot working ($0.5T_m < T < 0.75T_m$) or warm working ($0.3T_m < T < 0.5T_m$) or cold working ($T < 0.3T_m$) process, where T_m is the melting point of metal and T is the recrystallization temperature. But due to the competitive and increasing demand in the market, these conventional forming techniques are not sufficient. Therefore, the advanced forming techniques are adopted by industries to improve production and productivity. Among all the modern methods, high-speed forming (HSF) is old and developed for space components and widely used between the 1950s and 1960s. This method has come because superalloys and very complex geometry (thin-walled missile shells for the Apollo missions) could not be generated by conventional techniques. Although HSF has the initial interest, it failed to get a widespread acceptance around the 1970s. This was due to the lack of fundamental understanding of the science of each process and the behavior of material at high strain rates due to the limitation of tools and machines. But nowadays, HSF techniques are gaining its wide spread acceptance and most widely used in production (Dorel, 2007; Hingole, 2015). There are three major techniques of HSF that are described in further sections.

1.4.1 Explosive Forming

Shaping the parts in die by the application of explosives is known as explosive forming (Groover, 2018). The explosion generates shock waves in a fluid medium due to the chemical reaction that develops the forming pressure, by which the parts obtain their desired shapes. It is further classified into two groups according to the position of explosives, namely, standoff method and contact method.

1.4.1.1 Standoff Method

In this method, the metal plate is placed over the die that consists of the supreme space. This space is evacuated by a vacuum pump. Then, the whole assembly is kept inside the water, and the explosive is placed at an optimum distance from the plate. It is important to note that a segmented die can be used for complex shapes. This method is widely used to develop the rocket engine nozzle and space shuttle skin for space application, spherical tank, etc.

1.4.1.2 Contact Method

In this method, the explosive is kept in direct contact with the metal plate. The detonation of explosive creates a large interfacial pressure on metal by which it takes a desired shape. This method is widely used to create the bulging of tubes.

Advantages
1. Precise dimension is produced, eliminating the need of costly welding.
2. Large parts up to 4-inch square and 10-inch diameter can be efficiently formed.
3. It requires a low tooling cost.

Disadvantages
1. Labor cost is higher. Further, it is suitable for low production.
2. It cannot be performed in the open air.

1.4.2 Electro-Hydraulic Forming (EHF)

It is also known as electro-discharge or electro-shape or electro-spark forming process. In this process, a large amount of energy is applied in a very short interval by the application of higher voltage discharge. The applied higher voltage to electrodes creates the electrical breakdown that leads to the formation of a stable plasma channel, due to which a shock wave propagates and flows toward the metal blank through liquid medium. The force produced by shock wave is transferred to the forming surface of metal blank to get a desired shape. Applying external restraints in die or optimum amount of energy released can control the deformation in metal. It is important to control the process parameters such as standoff distance, size of the capacitor, type of transfer medium (generally water is used), vacuum and ductility of metal (it should be low). The materials such as Ni-alloy, stainless steel, Ti-alloy, Inconal-718 and Al-alloy can be formed using the EHF process. This process can be used to develop the radar disc, cone and other thinner shapes, and doors of passenger car, and miniaturize the complicated profile for the electronic industry (Zhang et al., 2018).

Advantages
1. This process is very efficient and cost-effective to produce a hollow shape compared with other forming techniques.
2. The desired shape is produced by a single step, and it enables deep forming. Further, it can easily form the fine and sharp line.
3. It does not depend on the electrical property of metal blank.

Disadvantages
1. It is suitable for small work. Further, the optimum standoff distance is necessary for an efficient process.
2. The equipment is costly and more complicated due to the requirement of vacuum.
3. It may create the surface defect by vaporization of electrodes that is because of impurities in fluid due to the higher voltage discharge.

1.4.3 Electromagnetic Forming (EMF)

It is also known as magnetic pulse forming. This forming process works on the principle of the Lorentz force. It states that the electromagnetic field of induced current is always opposing the electromagnetic field of inducing current. In this process, a large amount of energy is discharged in the form of current by a capacitor that travels through the coil conductor and produced a magnetic field. Further, when the coil is placed around a cylinder or within a conductive cylinder or near the flat metal sheet, it induces a secondary current. This inducing current is opposed to the induced current that results in the higher magnetic pressure. This pressure is moved toward the metal due to which the metal deforms into a desired shape. It is important to note that the direction of the movement of magnetic field is always away from the inductor coils. According to the interaction between tool and workpiece, the three categories used in EMF are electromagnetic compression to form a tube-and-hollow profile by enclosing of inductor in workpiece, electromagnetic expansion to form bulging in tube-and-hollow profile by application of inductor within workpiece, and electromagnetic sheet formation by positioning inductor in close proximity of the preformed part. There are various process parameters such as thickness of part, electrical conductivity (it should be higher), frequency (higher frequency is required for low conductive material or small thickness part), size of capacitor, gap between workpiece and coil (it should be smaller because small gap produces higher magnetic field and pressure) and the number of turns on coil that need to be controlled for efficient forming. This process is used to fabricate bulging in thin tubes, crimping of coils, tubes and wires, and bending of tubes into complicated shapes. Further, it has found an extensive application of a noncircular and hollow shape from the tubular blank. It can also be used to perform shearing, piercing and riveting operations (Gayakwad et al., 2014; Groover, 2018).

Advantages
1. This process is suitable for smaller tubes. Further, it can easily perform the operations such as collapsing, bending and crimping.
2. The process is precisely controlled by electrical energy.
3. It is much safer than explosive forming.

Disadvantages
1. This process is only applicable for electrically conductive materials. Further, it is not suitable for larger parts.
2. It is very critical to rigidly clamp the primary coil. Further, the life of the coil is reduced due to the large forces applied to it.

1.4.4 Peen Forming

This process is used to create a curvature on the thin metal sheet by application of shot peening (Brickwood, 1995). The shot peening has carried out either by using cast iron or steel balls of approximately 2.5 mm diameter that are discharged by a rotating wheel or by an air blast from a nozzle at a velocity of 60 m/s, due to which the surface of the metal sheet is experienced a compressive stress that causes the expansion of surface layer, resulting in the development of a curvature on sheet. This process also induces compressive residual stresses that improve the fatigue strength of sheet. Therefore, this process is widely used in the aircraft industry to produce complex and smooth curvature on the skin of aircraft wing. It is important to note that the larger diameter of 6 mm balls is used to produce the curvature on heavy section. Further, this process can also be used to produce straightening, twist or bent on parts (Groover, 2018).

Advantages
1. Low tooling cost. Further, the maintenance cost of tooling is negligible.
2. It can easily produce compound curvatures on sheet.
3. It improves the fatigue strength and stress corrosion resistance of sheet.
4. It requires a minimal lead time because of die-less forming.

Disadvantages
1. Limited forming potentials due to shallow deformation.
2. Required longer time to form the desired shape. Additionally, it requires another device to force out metal shots.

1.4.5 Superplastic Forming

This process works on the theory of super plasticity. It means the metal will go a large uniform elongation before necking and fracture. This process is used to create very complicated shapes and integrated structures, which are frequently stronger and lighter than the assemblies they have to replace. It is carried out at a higher temperature and controlled strain rate within the sealed die due to which it is produced a larger elongation compared with the conventional room temperature process (Giuliano, 2011). It is important to note that the strain rate has been controlled by the controlled flow rate of inert gases that is produced by the gas pressure on sheet, causes the elongation of sheet and fills the die cavity. This process is widely used to deform the materials such as Ti-alloy, stainless steel, Al-alloy, Al–Li alloy, Zn–Al alloy, and Bi–Sn alloy. Further, it is used to construct the fuel tank and muddy-guard of motorcars. It can also be used in critical aerospace applications where reduced weight is required.

Advantages
1. It produces parts with a better surface finish and precision.
2. Tooling cost is less. Further, it requires lower strength.

Disadvantages
1. The forming rate is very slow making it appropriate for lower production. Further, the production cost is higher.
2. At lower temperatures, the materials should not be superplastic.
3. This process can be used for restricted die design.

1.4.6 Thixo Forming

"Thixo" is a general term used to describe the near-net shape forming process from partially melted non-dendritic alloys within the die. When the metal is shaped in the open die, it is known as the thixo forming process. The metal shows rheological behavior and is carried out at a temperature between solidus and liquidus ranges. It produces minimal shrinkage, porosity and micro-segregation, but it requires high production cost due to lift over of non-recyclable billet.

1.4.7 Ceramic Forming

The ceramic forming is carried out by various processes such as slip casting, extrusion, injection molding, dry pressing, hot pressing, jiggering and isolated pressing. Among all, slip casting is the most commonly used process to form the ceramic product. It is also known as drain casting (Groover, 2018). The basic steps involved in slip casting are as follows:

1. First, the ceramic particles are crushed into small size using ball mill, and then it is blended with additives and immiscible liquid. The prepared colloidal suspension is known as slip.
2. The slip is poured into the mold typically made by POP. The slip is easily flowing inside the mold due to its excellent fluidity and optimum viscosity.
3. After pouring of slip, the outer layer of mold absorbs the small amount of suspension and then drains it out after the generation of suspension layer on mold.
4. Finally, the dried part is removed from mold, which is known as green part followed by heat treatment to improve the part strength.

It is important to note that the slip is continuously supplied inside the mold to reduce the shrinkage for the solid ceramic part. The obtained part is known as semi-rigid or soft solid. Another important point to note is that if any decorative design or machining are needed on the ceramic part, then it is performed on the green part followed by heat treatment. Slip casting can be widely used to develop complex and large parts such as plumbing ware, art objects and dinnerware. Further, although the cost of equipment is lesser, it shows a lower production rate and poor dimensional accuracy.

Moreover, the **doctor-blade process** is used to fabricate the ceramic sheet of <1.5 mm thick. In this process, the slip is poured over a moving plastic belt and thickness has been controlled by the blade that is attached inside the chamber. The typical ceramic products are electrical insulators, seals, ovenware, tiles, ball and roller bearings, rotor for gas turbine and lightweight components for high-speed machining.

1.4.8 Glass Forming

Glass is formed through melting followed by shaping in mold using tool or blow. In all glass-forming and -shaping processes, the molten glass is heated up to 1000°C–1200°C, which causes the red-hot viscous syrup that is supplied from the furnace or tank. The glass products are categorized into four types. These are flat sheet or plates, rods and tubes, discrete products like bottles, bases, headlights, and television tubes, and glass fiber-reinforced composites for fiber optics (Groover, 2018).

The **flat sheet and plates** are manufactured by three methods that are briefly described below:

1. **Float method:** In this method, the molten glass is fed from the furnace to long float bath of molten tin under controlled atmosphere and at 1150°C temperature. Further, the glass is moved at about 650°C temperature to another chamber through rollers. Here, the glass is solidified and shown a very smooth, so further polishing or grinding is not required.
2. **Drawing process:** In this method, the molten glass is squeezed between forming rollers and turning rollers and passed through the pair of rollers that are arranged in a similar old-fashioned clothes wringer.
3. **Rolling process:** In this method, the molten glass is squeezed between power rollers and then formed a glass sheet. It is important to note that the texture on the glass sheet has been produced by using the textured rollers means the glass surface is the replica of roller surface. Therefore, the surface of glass sheet produced by this method is appeared rough compared with the float method.

Further, **tubes and rods** are manufactured by using a hollow mandrel. The molten glass is wrapped around the hollow mandrel and then drawn through a set of rollers. To create the hollow tube and rod of glass, the air is blown through a hollow mandrel. This method is widely used to develop glass tubes for fluorescent bulbs. However, the **discrete glass products** are developed by four methods:

1. **Blowing:** It is also known as the blow-molding process. The air is blown on a heated glass against the inner wall of mold. To prevent the sticking of glass from mold surface, oil or emulsion is applied. In this process, it is difficult to control the wall thickness, but it can be used for higher production of glass components.
2. **Pressing:** In this process, the molten glass is kept into a mold and pressed by a plunger in the confined area. This process is similar to close die forging. It is important to note that the pressing is not carried out in one-piece mold if the product has thin wall and the shapes through which the plunger is not easily retracted. In this case, the split mold is used. For example, split mold is used in the fabrication of bottles.
3. **Centrifugal casting:** It is also known as the spinning process. The working principle is similar to centrifugal casting for metals. The mold is rotated at a higher speed that applied a centrifugal force on molten glass, causes the pushing of molten glass alongside the mold wall. Typical products made by this process are large lenses for research telescopes, dishes, sunglass lenses, mirrors for telescopes and lighting panels.
4. **Sagging:** In this process, the glass is kept over the mold and starts heating it. Due to heating and own weight of glass, it starts to sag and takes the shape of the mold. This process is similar to the thermoforming of thermoplastic without pressure. It can also be used to fabricate the dishes, sunglass lenses, mirrors for telescopes and lighting panels.

1.4.9 Incremental Sheet Metal Forming

This process is performed without the die with the help of a numerically controlled tool on a rigidly clamped sheet. Therefore, this process is also known as the die-less forming process. The tool is applied localized pressure due to which local deformation occurs on the sheet metal. Its collective effect gives the desired geometry. It is a very economical process for low production with better surface quality and applies lesser forming pressure. The process parameters such as tool size, incremental depth, rotational speed of tool, feed rate, forming angle, thickness of sheet and lubricant between tool and workpiece are important to be considered for efficient processing. This process is widely used to fabricate the various sheet metal components in aerospace and automobile industries such as body panels, indoor and outdoor panels, instrument panels, engine cover and passenger sheet cover (Cao et al., 2008; Ajay et al., 2019). This process is divided into two categories, i.e., single-point incremental forming (SPIF) and two-point incremental forming (TPIF). The TPIF is again divided into three categories. These are with partial die support, with full die support and with supporting tool.

The SPIF process is also known as negative incremental forming. In this process, the forming tool applies the pressure on a rigidly clamped sheet or blank to get a desired shape without any addition of die or counter tool. It is a very simple process due to which it is widely used for the fabrication of prototypes that does not require the part to acquire intricate tolerances. This process is highly flexible and economical for small batch production, and it requires minimum forming forces. This process cannot control the uniform wall thickness and give the lower accuracy in profile due to the spring back effect.

TPIF is also known as positive incremental forming. It requires the die or counter tool to get the desired deformation on sheet. In the TPIF process, two points are in contact. These are the contact between the sheets and forming tool and the contact between the sheet and supporting tool (either die or counter tool). TPIF gives the better forming characteristic from SPIF.

1.5 ADVANCED JOINING TECHNIQUES

The joining process is very old. It started from ~1000 BC in ancient civilizations after the discovery of Cu, Ag, Au, Fe and bronze. The first joined part is thought to be gold ornament that was developed by Egyptians. The joining processes were mostly the same up to the first industrial revolution, but it was transformed during the 19th century. In the 19th century, industries added carbon arc welding, resistance welding and oxyfuel welding for joining materials. In addition, joining processes were further transformed in the 20th century after World Wars I and II. In this era, industries further developed arc welding to include gas tungsten arc welding, submerged arc welding, metal inert gas welding and plasma arc welding. Developments were also made in resistance welding such as seamless welding and projection welding. Further, at the start of the 21st century, magnetic pulse welding was introduced by Pulsar Ltd. of Israel. It used capacitive power as an energy source to join the metal. Further, microwave welding is another addition to the 21st century (Mehta, 2017). In any joining process, the energy source is very important. Therefore, Table 1.1 shows the various types of welding processes with energy sources that are used to join the metal. Further, few advanced joining processes such as magnetic pulse welding, friction stir welding and explosive welding from solid-state welding and laser beam welding, electron beam welding and microwave welding from fusion welding are described briefly in the following sections.

TABLE 1.1 Summary of various welding processes with their energy source for joining

SOLID-STATE WELDING (WITHOUT FILLER METAL, MECHANISM: PLASTIC DEFORMATION WITH/WITHOUT RECRYSTALLIZATION)	
TYPES OF WELDING	ENERGY SOURCE
Super plastic forming, explosion welding, cold welding, forge welding, friction welding, ultrasonic welding, hot pressure welding, roll welding, friction stir welding, explosion welding, diffusion welding, creep isostatic pressure welding	Mechanical
Stud arc welding, magnetically impelled arc butt welding, resistance spot welding, resistance seam welding, projection welding, flash welding, upset welding, percussion welding, resistance diffusion welding, magnetic pulse welding	Electrical and mechanical
Pressure gas welding, exothermic pressure welding, pressure thermite forge welding	Chemical and mechanical
Fusion Welding (with Filler Metal, Mechanism: Meting and Solidification)	
Arc welding, gas welding, gas metal arc welding, shielded metal arc welding, submerged arc welding, electrogas welding, electroslag welding, flux cored arc welding	Consumable electrode
Gas tungsten arc welding, plasma arc welding, carbon arc welding, atomic hydrogen welding, stud arc welding	Non-consumable electrode
Resistance spot, resistance seam, projection welding, flash/upset welding, percussion, induction welding	Resistance
Laser beam welding, electron beam welding, infrared welding/brazing, imaging arc welding, microwave welding	Radiant energy

1.5.1 Magnetic Pulse Welding

It works on the principle of Lorentz force meaning the electromagnetic field of induced current is always opposing the electromagnetic field of inducing current. In this process, a large amount of energy is discharged in the form of current by primary circuit (capacitor bank) that travels through the coil conductor and produces a magnetic field. Further, when workpiece is placed in the vicinity of coil, then it induces a secondary current. This inducing current is opposed to the induced current that results in the higher magnetic pressure. This pressure generates the larger stresses that are higher than material flow stresses, which causes the plastic deformation in workpiece and then accelerates joining of workpiece. It is used only to join the conductive metal such as Al, Brass or Cu to steel, Ti, Mg-Cu in a lap joint. It is a solid-state welding process similar to explosive welding. There are various process parameters such as the thickness of workpiece, electrical conductivity (it should be higher), frequency (higher frequency is required for low conductive material or small thickness part), size of capacitor, gap between workpiece and coil (it should be smaller because small gap produces a higher magnetic field and pressure) and the number of turns on coil needs to be controlled for efficient welding. Initially, it was developed for nuclear energy application. Further, it is successfully used in automotive, aerospace, ordnance, consumer products, packaging and electrical industry (Sapanathan et al., 2016).

Advantages
1. It can be used to join metallic or non-metallic materials within a short setup time.
2. It is suitable for high production and reduces the cost of manufacturing of components.
3. It is a cold joining process. Therefore, no heat-affected zone and no internal stresses are developed inside the material.
4. The welding joints are stronger and lighter.

Disadvantages
1. It is only used for conductive materials. For non-conducting materials, the conducting plate is used.
2. It is difficult to optimize the gap between workpiece that is important to develop the correct impact speed.
3. It is not suitable to join the larger workpiece.

1.5.2 Friction Stir Welding

It is another new solid-state welding process. The basic principle of this welding is the same as the friction welding process (Yuce et al., 2015). In this process, the heat is generated at the interface due to the friction and then metal diffusion is started at the interface. At this point, the high-pressure force is applied which accelerates the diffusion and then performs the metal to metal joint. Similarly, in friction stir welding, the pressure and friction at the interfacial surface are applied through a rotating tool that is rotated on its own axis and moved longitudinally at the interfacial surface of plates. This movement has generated the deformation and produced the heat at the interfacial surface because of friction between tool and material, causes the diffusion and then pressure force is applied through a tool on diffused portion followed by joining of materials. The basic mechanism to join the two materials is thermo-mechanical treatment at the interface. The most important process parameters in friction stir welding are rotational speed of tool, tool design and tool tilt, plunge depth, transverse speed of tool and applied force on tool. It is most widely used to join the Al-alloy by using a high-speed steel tool. It is important to note that tungsten and iridium tool is used for joining of steel and tungsten alloys. Further, this process is used in various industries such as aircraft industry (to join the wing, fuel tank, aircraft structure, etc.), marine industry for structural work, automobile industry (to join the fuel tank, wheel rims, chassis, etc.), chemical industry (pipelines, heat exchanger, air conditioner, etc.) and electronic industry (to join the bus bar, Al–Cu, etc.).

Advantages
1. It is a very efficient process. Further, it can be operated in all directions.
2. It produces low distortion and shrinkage. Further, it does not produce arc and porosity.
3. It provides excellent tensile, fatigue and bend strength.

Disadvantages
1. The workpiece must be rigidly clamped. Further, it cannot weld the non-forgeable materials.
2. It provides the keyhole at the end of each weld.
3. Compared to fusion welding, the initial cost of equipment is high.

1.5.3 Explosive Welding

It is another important solid-state welding process that is used to join two similar or dissimilar materials (Lysak and Kuzmin, 2011). It is categorized into two configurations, i.e., oblique and parallel. The basic difference between these two configurations is the thickness and size of the plate. If the thickness and size are small, then oblique configuration is used, while if the thickness and size are large, then parallel configuration is used. In both configurations, the arrangements are the same except the collision angle between the parent material and the flyer plate (in oblique, it is $5°C–25°C$, and in parallel, it is $0°C$). The arrangement in both configurations is parent material → flyer plate → buffer layer → explosive layer → detonator. This process works on the basic principle of metallurgical bonding. The controlled detonation of explosives is used to join the two materials. The higher pressure (10–100 kbar) at the flyer plate is developed due to the controlled detonation of explosives at higher velocity (1000–7000 m/s). The higher pressure generates the large compressive forces on the interfacial surfaces within a short duration of time that causes melting and diffusion at the interfacial surface due to which it forms the metallurgical bond and joins the two surfaces. This bond is stronger than the parent material. The process parameters that need to be optimized before the welding are standoff distance, collision angle, detonation velocity, surface preparation and type of explosive (RDX, TNT, lead azide, PETN, etc.). This process is capable to join the large surface area due to the distribution ability of higher heat density that is produced by controlled detonation of explosives. This process is most widely used to join the Al-stainless steel. Further, it is used to join the cooling fan, pipe, concentric cylinder, and tube and also used to join the dissimilar materials that are not possible to join by other welding processes.

Advantages
1. It can be used to join both similar and dissimilar materials. Further, the large surface can be joined in a single pass.
2. It does not affect the properties of welding materials.
3. It shows a higher production rate.

Disadvantages
1. Due to the explosion, it creates more noise. Further, the welding is highly dependent on the process parameters.
2. It can only weld ductile materials with higher toughness.
3. It can only be used for limited designs.

1.5.4 Electron Beam Welding

In this process, a cup-shaped tungsten electrode is used to discharge the electrons toward the anode that is having the holes. The intensity of electrons is increased by the larger difference in the applied voltage. The discharge electrons are created in the magnetic field with the help of a field electrode that is further converted into an electron beam. The developed higher energy electron beam hits the workpiece material due to which the surface of material is melted and performs the joining process. This process can produce very deep penetration and good quality weld with insignificant heat-affected zones. It is important to note that when the electron beam hits the workpiece then X-rays are developed due to which the whole welding setup is kept inside the vacuum chamber that consists of the lead lines for absorbing these developed X-rays. It can be used to join the Al and its alloy, Mg alloy, titanium, tungsten, gold, Cu-steel, bronze-steel, Cu and its alloy, and ceramics (electrically conductive). This process is widely used in the space industry, automobile industry, tool construction and power plants (Kalpakjian, 2014).

Advantages
1. It produces an insignificant heat-affected zone. Therefore, it produces a high-quality weld.
2. This process can join the thin (0.1 mm) and thick (300 mm) plates. Further, it can be used to join the extreme narrow grooves (t:b→50:1)
3. The welding can be performed at a very high speed.

Disadvantages
1. Higher equipment and maintenance cost is required.
2. It requires higher vacuum inside the chamber to perform the welding. Further, the electron beam may be deflected due to magnetism.

1.5.5 Laser Beam Welding

In this process, the atoms are pumped from ground state to higher state by using a xenon lamp. When the populations of atoms are reached sufficiently in higher state, these atoms are moved to metastable state and further these atoms are moved to ground state. The movement of atoms from metastable state to ground state is emitted photons. The higher population of photons is directed to form a beam. This beam is known as a laser beam that is utilized in the joining process through melting of the metal surface. It is important to note that one side of the lamp is 100% reflecting and another side is 98% reflecting. It means that the photons are not allowed to come out until and unless it will not come in a particular direction. Further, due to the higher population of photons, they are moved to and fro inside the tube, and then they lose their energy that is carried out by a cooling agent. Thus, the laser efficiency is very low. Moreover, laser beam welding is more versatile as compared with electron beam welding because here, material is welded in air that can be observed through a transparent medium. It can also be used to perform profile welding. Therefore, this process is widely used in electronic and aerospace industries for precision welding. For joining of aluminum, the Ne-YAG (neodymium yttrium aluminum garnet) laser is used. The important parameters that affect the efficiency of process are medium, atmosphere, and power of laser beam, beam intensity, beam diameter and melting temperature (Kalpakjian, 2014; Mehta, 2017).

Advantages
1. In this process, the filler is not required and can be used to join two dissimilar materials.
2. It is used in the open air. Further, it can be used to join superalloy metal without difficulty.
3. It produces an insignificant heat-affected zone. Therefore, it produces a high-quality weld.

Disadvantages
1. Higher equipment and maintenance cost is required.
2. It can produce cracks in some materials due to the rapid cooling rate.

1.5.6 Microwave Welding

Microwave welding is a form of electromagnetic welding that uses a radiation frequency of typically 2.5 GHz (Metaxas and Meredith, 1983). The higher frequency causes localized heating and melting followed by diffusion of materials and welding. Heating is produced through various physical processes such as polarization heating (either by electrons or by ions), electrical resistance heating and heating due to the Maxwell–Wagner effect. It is divided into two categories, i.e., with auxiliary materials (also known as indirect heating) and without auxiliary materials (also known as direct heating). Direct heating is used to join the components made of metal that are heated up in the microwave field and indirect heating is more feasible to join the thermoplastics but it is appropriate for microwave inactive plastic, which cannot heat directly.

In the indirect heating process, a third material is inserted between two joining parts and started heated up in the microwave field. Due to heating, it is diffused in weld beads and joins the two parts (Wise and Froment, 2001). Input frequency, input power, dielectric property and interface temperature are important parameters to optimize the efficient microwave welding process. It is a very versatile technique to join the two components in the automotive industries. The lower capital cost of equipment, low power consumption and faster heating are the advantages of this welding process. Further, it has only disadvantage that it is required a consumable microwave disposed materials at the joining interface.

1.6 ADVANCED COMPOSITES MANUFACTURING TECHNIQUES

Mesopotamians developed the first composite material in 3400 BC. They glued the strips of wood at various angles. Further, Mesopotamians and Egyptians developed more composites by the addition of straw in bricks, pottery and boats. However, in 1200 AD, Mongols developed the first composite bow by a combination of wood, bamboo, bone and silk in natural pine resin. These weapons were further developed more accurately and precisely until the 14th century. The revolution in composite manufacturing was initiated after the development of synthetic polymers such as Bakelite, melamine and celluloid in the 19th century. In the 20th century, it was further improved after the development of plastics such as polystyrene, phenolic and polyester. But it was recognized more in 1935 after the development of fiber-reinforced polymer and it was used at the time of World War II. Then slowly the market of composites increased and developed the carbon fiber that was patented in 1960. This development has improved the stiffness to weight ratio of composites. Around 1980, the development of ultra-high molecular weight polyethylene was added in composites and made them suitable for aerospace components, structural and personal armor, sporting equipment, medical devices, etc. In the beginning of the 21st century, the composite arrived into the mainstream of manufacturing and construction (Campbell Jr., 2003). Thereafter, industries were introduced to the MMC and ceramic matrix composite (CMC) in this family. Nowadays, industries are more focused on the manufacturing of nanocomposites. These composites consist of two or more phases in which one phase should have the dimension in the range of nanometers. These phases may be organic–organic, inorganic–inorganic or organic–inorganic; those are blended in different matrix materials such as ceramic, metal and polymer. The nanocomposites exhibit excellent mechanical, electrical, optical and structural and electro-chemical properties from traditional composites (Campbell Jr., 2003; Groover, 2018). Therefore, it is very important to understand the advanced composite manufacturing methods for PMC, MMC and CMC that are discussed briefly in the following subsections.

1.6.1 Fabrication Techniques for Polymer-Based Nanocomposites

When polymer is used as a matrix, that composite is known as PMC. The polymer is categorized into four groups that are linear polymer (chain molecules), thermosetting polymer (highly cross-linked molecules), thermoplastic polymer (molecules are not interconnected) and elastomer (thermoplastic or lightly cross-linked thermosets, having elastic deformation >200%). The crosslinking of molecules in polymers is occurred due to the polymerization process. The degree of polymerization is equal to the number of monomer units in the chain. Among all types of polymers, only thermosetting and thermoplastic polymers are widely used for the manufacturing of nanocomposites. All methods are summarized in Table 1.2.

1.6.2 Fabrication Techniques for Metal-Based Nanocomposites

If metal as a matrix and ceramic or metal as reinforcement is used, then it is known as MMC, e.g., Al-MMC, Mg-MMC, Ti-MMC and Cu-MMC. There are three major methods to develop an MMC. These are liquid-state fabrication, solid-state fabrication and in situ fabrication (Campbell Jr., 2003; Groover, 2018).

1.6.2.1 Liquid-State Fabrication of MMC

In this process, the dispersed phase is blended in a molten matrix material using different techniques such as stir casting, squeeze casting, infiltration and deposition followed by its solidification. Stir casting and squeeze casting were described in Sections 1.2.11 and 1.2.7, respectively.

1.6.2.1.1 Infiltration
This is the liquid-state fabrication method for MMC. In this process, the preform of disperse phase is shocked in a molten material that is filled the

TABLE 1.2 Fabrication process of polymer-based nanocomposites with their advantages, disadvantages and applications

| | PROCESSES FOR DEVELOPMENT OF THERMOSETTING MATRIX NANOCOMPOSITES | | |
METHODS	PROCESS	ADVANTAGES	DISADVANTAGES	APPLICATIONS
Hand lay-up	Performed in open mold with the application of gel coating followed by fiber layer	Simple and easy, low tooling cost, large parts can be produced	Slow, labor-consuming job, not good quality and strength	Structural applications
Spray-up	Resin and chopped fibers are sprayed through two distinct sprays on open mold	Lower labor cost, design flexibility, low tooling cost	Slow, labor-consuming job, longer curing time, high waste factor	Caravan bodies, truck fairings, bathtubs, small boats
Filament winding	Resin-impregnated fibers wound over a rotating mandrel at the desired angle	Low scrap rate, formed non-cylindrical parts, flexible mandrel	Control the filament tension, chosen carefully viscosity and pot life	Pressure vessels, storage tanks and pipes, drive shafts, rocket motors, launch tubes
Pultrusion	Resin-impregnated fibers are pulled from the heated die to make a part	Simple, low-cost and high-volume production process, good surface quality	Die jamming, fiber breakage, improper fiber wet-out	Electrical insulators, panels, beams, gratings, ladders

(Continued)

TABLE 1.2 (CONTINUED) Fabrication process of polymer-based nanocomposites with their advantages, disadvantages and applications

	PROCESSES FOR DEVELOPMENT OF THERMOSETTING MATRIX NANOCOMPOSITES		
METHODS	PROCESS	ADVANTAGES	DISADVANTAGES	APPLICATIONS
---	---	---	---	---
Pulforming	Similar process as pultrusion with length formation in semicircular contour	Simple, low-cost and high-volume production process, good surface quality	Die jamming, fiber breakage, improper fiber wet-out	Curved cross-sections
Resin transfer molding	Inserted preformed and oriented reinforcement in the heated close mold die, then poured the resin and make the part	Easily made very large and curvature parts, less time, high-volume production process	Complex mold design, fiber may wash or move during resin transfer	Automobiles, aerospace, sporting goods and consumer products
Autoclave molding	Closed vessel process under simultaneous application of high temperature and pressure	Less voids, excellent properties, cured many parts at the same time	Expensive	Aerospace industry to fabricate high strength/weight ratio parts
Compression molding	Heated thermosets are kept in lower die and apply the pressure through punch to get the desired shape	Low cost, fast setup time, mold heavy plastics	Lower production rate, reject part cannot be reprocessed	Brush, mirror, handle, trays, cookware, automotive parts

(Continued)

TABLE 1.2 (CONTINUED) Fabrication process of polymer-based nanocomposites with their advantages, disadvantages and applications

METHODS	PROCESS	ADVANTAGES	DISADVANTAGES	APPLICATIONS
	PROCESSES FOR DEVELOPMENT OF THERMOSETTING MATRIX NANOCOMPOSITES			
Processes for Development of Thermoplastic Matrix Nanocomposites				
Injection molding	Polymer and fiber mixture are melted and move forward through reciprocating screw in mold cavity	Fully automated, highly productive, highly accurate, easily fabricate complex shape parts	Higher tooling cost, limited length of fibers decreasing their reinforcing effect	Boat hulls and lawn chairs, to bottle cups. Car parts, TV and computer housings
Diaphragm forming	Insert thermopreg fabric in between heating and forming silicon sheets	Formed double-curvature, easy fabrication		Engine cover, double-curvature components
Automated lay-up	Layers of prepreg (reinforcing phase impregnated by liquid resin) tape are applied on the mold surface by a tape application robot	Low cost than hand lay-up	Limited to flat or low curvature	Airframe components, bodies of boats, truck, tanks, swimming pools and ducts

(Campbell Jr., 2003; Groover, 2018)

space available in preform. This process can be performed either naturally using capillary force or forcefully using gases or mechanical force. When the process is performed using external forces, it is known as forced infiltration. It is noticed that low damage of fibers is reported in gas infiltration than in mechanical infiltration. This method is widely used to fabricate W-Cu MMC.

1.6.2.2 Solid-State Fabrication of MMC

In this process, the bonding between matrix and reinforcement is obtained through mutual diffusion under high pressure and at elevated temperatures. The solid-state fabrication methods are classified into two categories. These are sintering and diffusion bonding.

In the **diffusion bonding process**, metal foils (matrix) are stacked in a specific direction with long fibers as a dispersed phase and then higher pressure is applied at elevated temperature to develop an MMC. This process is used to simplify the shape of parts like plates, tubes, etc.

In the **sintering process**, the green compact (made by blending powders in a ball mill followed by compression molding) is heated at elevated temperatures below the melting point of metal. Due to heating, the powder particles diffuse each other and create a strong bond. This process is widely used in powder metallurgy components for various applications in the automobile industry.

1.6.2.3 In Situ Fabrication of MMC

In this process, the dispersed phase is developed inside the matrix as a result of precipitation from the melt during its solidification. This process provides the more uniform dispersed phase particles, but the choice of dispersed phase particles is limited due to their thermodynamic ability of precipitation in a particular matrix.

1.6.2.3.1 Deposition

This method is widely used to develop the MMC coating for very harsh applications. The deposition has performed by using electrolyte or spray or vapor methods (Campbell Jr., 2003; Groover, 2018).

In **electrolyte deposition**, the electrolyte solution of metal matrix ion with dispersed particles phase is deposited on a component as a coating. For example, $Ni-Al_2O_3$ coating for oxidation resistance; Ni-SiC coating for wear resistance; and Ni-PTFE, Ni-C and $Ni-MoS_2$ for antifriction coating.

In **spray deposition**, thermal spraying techniques are used, in which atomized molten droplets with dispersed phases are sprayed using high-velocity gas stream on substrates. For example, WC-Co coating is performed using high-velocity oxy-fuel (HVOF).

In **vapor deposition**, the molten metal with dispersed phases is evaporated and condensed or react chemically through precursors on the substrate. These processes are physical vapor deposition and chemical vapor deposition.

1.6.3 Fabrication Techniques for Ceramic-Based Nanocomposites

If ceramic as a matrix and ceramic or metal as reinforcement is used, then it is known as CMC. These composites are designed for improved toughness, but the main limitation of composite is brittleness (Singh et al., 2017). The various methods used to develop the CMC are briefly described below.

1.6.3.1 Polymer Infiltration and Pyrolysis (PIP)

In this method, low-viscosity polymer is infiltrated in ceramic fiber preforms through capillary action and then performed pyrolysis at 800°C–1300°C in the environment of argon inert gas. This process decomposes the polymer and releases the volatile gases that are finally converted into a porous structure of CMC.

1.6.3.2 Chemical Vapor Infiltration (CVI)

In this process, reactant gases are diffused into a porous preform with long continuous fiber, and then deposition happens. This deposited material is reacted chemically with long continuous fibers and form the CMC. This process is similar to the chemical vapor deposition process where coating or deposition is developed through the chemical reaction of precursor gas to the component. It is widely used to fabricate the silicon carbide matrix composite reinforced with long continuous fibers of silicon carbide.

1.6.3.3 Liquid Silicon Infiltration (LSI)

It is one type of reactive melt infiltration process. In this process, the liquid metal of Si is infiltrated into porous carbon preform that is developed by either PIP or CVI process at 1400°C. The molten material of Si is shocked by carbon preform through capillary action. The shocked melt is reacted with the carbon preform and made SiC ceramic matrix. The composite developed by LSI is insignificant or zero residual porosity compared with CVI and PIP processes.

1.6.3.4 Direct Melt Oxidation (DIMOX)

It is another type of reactive melt infiltration process. In this process, the molten metal is infiltrated into porous preform in the oxygen gas environment. The molten material is shocked through capillary action by preform and forms a thin oxide layer front on preform through the chemical reaction with oxygen. This reaction grows the ceramic matrix layer and forms a CMC. The developed composite is not having any pores and impurities that are usually present in the case of ceramic fabrication through the sintering process. This method is basically used for the development of composites with ceramic matrix from aluminum dioxide.

1.6.3.5 Sol-Gel Infiltration

In this method, the low-viscosity liquid colloidal suspension of ceramic particles, known as sol, is infiltrated into preform. The liquid colloid is shocked by preform through capillary action at elevated temperatures. Due to high temperatures, sol is converted into solid gel by polymerization or hydrolysis mechanism. It is important to note that the quantity of ceramic gel is low due to which significant shrinkage is occurred after drying. Further, repeating the infiltration process enhances the densification of CMC.

1.6.3.6 Slurry Infiltration

In this method, the ceramic particles slurry is dispersed into a liquid carrier that consists of additives and binders. Then, the developed slurry is infiltrated into a porous preform. The slurry is shocked by porous preform through a capillary action. After that, the preform is dried and performed hot pressing to develop a CMC. This process is used to develop the ceramic–glass matrix, ceramic matrix and glass matrix. This method is similar to the sol-gel infiltration method but it produces a denser structure with smaller shrinkages of CMC.

1.7 ADVANCED MACHINING TECHNIQUES

Machining also has a very long history. It was started in Egypt in 1200 BC. They handcrafted the wood by using a lathe machine tool. It was indicated that machining has played a very important role in civilization. As time moves, various types of specific machine tools were developed to fulfill the needs of humans. For example, John Wilkinson developed the boring mill to fabricate the bored cylinder. Further, at the time of the first industrial revolution, the waterpower machine tool unsealed a new door for enhanced machining. Moreover, in the 19th century, the demand for more refined machine tools increased that has created a path for the new development of machine tools such as drill press, milling, lathe, shaping, grinding and planning and is now readily available in all machine shops for large-scale production. However, in the 20th century, the demand for miniaturized components increased due to which industries have added further computer numerically controlled machine tools in their machine shops. The more precise machining of miniaturized components up to the micro level has been started in the 21st century for the compact design of components. These precise machining technologies are known as advanced machining techniques that are summarized in Table 1.3. It shows the classification of advanced machining techniques associated with their types of energy and source of energy used, principle of material removal takes place and potential uses (Ghosh and Mallik, 2010; Payal and Sethi, 2003; Jain, 2010; Lee et al., 2016).

USMM, ultrasonic micromachining; AJMM, abrasive jet micromachining; WJMM, water jet micromachining; AWJMM, abrasive water jet micromachining; EDMM, electrical discharge micromachining; EBMM, electrical discharge micromachining; LBMM, laser beam micromachining; LAEDM, laser-assisted electrical discharge machining; IBM, ion beam machining; PAM, plasma arc machining; CHM, chemical machining; ECMM, electrochemical micromachining; ECSM, electro-chemical spark machining; CMP, chemo-mechanical polishing; ELID, electrolytic in-process dressing.

1.7.1 Thermal-Assisted Techniques

These techniques are widely used to machine those materials whose machinability index is very high such as superalloys and Inconal-718 (Rahim et al., 2015). These materials are difficult to machine by any conventional machining process. Therefore, advanced machining processes are used in which thermal-assisted machining techniques are used widely. This technique uses the localize melting due to which the surface of material is softened locally that is reduced yield strength, strain hardening and hardness that further changes the deformation behavior from brittle to ductile. Therefore, the basic mechanism of material removal is melting and vaporization. These are various heat energy sources such as plasma, laser, electron beam, spark and induction heating for localized melting is used for machining. Further, it also reduces the machining issues such as cutting forces, tool life, surface integrity and change of mechanical properties of workpiece.

1.7.2 Plasma-Assisted Machining

In this process, the inert gases are heated through an electric arc up to plasma state that is the fourth state of matter due to which it is converted into superheated and electrically ionized gases at 5000°C. That is directed toward the

TABLE 1.3 Classification of advanced finishing techniques

TYPES OF ENERGY	ADVANCED MACHINING TECHNIQUES	SOURCE OF ENERGY	PRINCIPLE OF MATERIAL REMOVAL	USES
Mechanical	USMM	Mechanical motion	Erosion of workpiece mechanically	Round and irregular holes, impressions, etc.
	AJMM	Pneumatic		Drilling, cutting, deburring, etc.
	WJMM	Hydraulic		Paint removal, cleaning, cutting frozen meat, etc.
	AWJMM	Hydraulic		Peening, cutting, textile, leather industry, etc.
Thermal	EDMM	Electric spark	Melting and evaporation of workpiece	Holes in nozzles and catheters, channel cutting, curved surfaces, etc.
	EBMM	High-speed electrons		Drilling fine holes, cutting contours in sheets, etc.
	LBMM	Powerful radiation		Drilling fine holes, cladding, etc.
	IBM	Ionized substance		3D patterning in ICs, micro tools and micro dies fabrication, etc.
	PAM	Ionized substance		Cutting plates, etc.

(Continued)

TABLE 1.3 (CONTINUED) Classification of advanced finishing techniques

TYPES OF ENERGY	ADVANCED MACHINING TECHNIQUES	SOURCE OF ENERGY	PRINCIPLE OF MATERIAL REMOVAL	USES
Chemical	CHM	Corrosive agent	Corrosive reaction	Pockets, contours, MEMS, etc.
Electro-chemical	ECMM	Electric current	Ion displacement	Blind holes, cavities, etc.
Electro-chemical and Thermal	ECSM	Electrical discharges	Melting and evaporation, and chemical etching	Holes, grooves, channels, complex shape contours, etc.
Chemical and mechanical	CMP	Abrasive slurry with chemicals	Chemical reaction and mechanical abrasion	Polishing of Au and Ti, optoelectronic components, etc.
Electro-chemical and mechanical	ELID	Electric current and mechanical motion	Electrolytic reaction and grinding action	Small hole grinding, fine finish of hard and brittle materials, etc.
Thermal	LAEDM	Powerful radiation and electric spark	Ablation	High-quality drilled holes and channels, etc.

materials through high-velocity stream, which is caused by the localized melting on the surface of material and changed the deformation behavior from brittle to ductile, which further vaporized the small amount of materials from surface (Jain, 2010). This process produces a very high degree of heat concentration due to which it is difficult to control precisely. This process can be used to machine the aluminum alloys, alloy steel, magnesium and stainless steel.

1.7.3 Laser-Assisted Machining

It is a hybrid machining process that pools the preheating and mechanical removal (Rahim et al., 2015; Jain, 2010). In this process, the controlled intensity of laser beam is directed on the workpiece surface for localized melting at ~700°C and 140 W power, which change the surface from brittle to ductile. After softening of materials, the cutting tool comes into contact and removes the unwanted material from surface to get a desired shape and geometry. The distance between the laser and cutting tool is a very important parameter apart from cutting speed, laser power, etc. This process can be combined with turning, milling and grinding as per the requirement. The main limitation of this process is the costlier setup and absorption rate on various materials.

1.7.4 Electron Beam Assisted Machining

It is also known as high-energy beam machining process (Rahim et al., 2015; Jain, 2010). In this process, the high-energy electrons are produced through an electron gun that are concentrated at anode due to electrostatic field and accelerated toward the workpiece. This provides the localized melting, causes softening of materials and further vaporized the small amount of material from surface. The whole process is performed under vacuum environment. There are various process parameters such as accelerating voltage, beam current, pulse duration, energy per pulse and vacuum that are needed to be optimized for efficient machining. It can be produced the smaller diameter of hole up to 0.02 in. It is widely used for machining of thin materials.

1.7.5 Ion Beam Machining

This process is very important in the age of nanotechnology for ultraprecise machining (Jain, 2010). In this process, a stream of energetic ion beam is bombarded on the surface of workpiece under the vacuum environment. These ions transferred their energy to surface atoms due to which the atom from the surface is removed because of excess of actual energy from binding energy. This process is known as sputtering. Ion current density and incident angle (i.e. max. ~80°C) are very important parameters in this process because if ion density is increased then the chance of irregular machining on surface is more. Further, a larger incident angle reduces the sputter yield due to which ion approaches the glancing incident. This process is widely used in the semiconductor industry for ultrafine machining.

1.8 ADVANCED FINISHING TECHNIQUES

Surface finish is widely used to enhance the surface properties of any components such as enhanced appearance, wettability, corrosion resistance and wear resistance, chemical resistance and electrical conductivity. It is always tough and difficult to precise finishing of internal surfaces and complex geometry. To attain the precise finishing, many advancements are taken place with the help of abrasives, having small multiple cutting edges that are used to remove the unwanted material from surface through shearing and cutting to attain the desired shape and surface finish up to the micron to nanometer range. The traditional finishing such as lapping, honing and buffing is worked on this principle. But these processes are difficult to finish the complex surfaces, free from surfaces and intricate areas of distinct surfaces. Therefore, advanced finishing processes are required to examine finishing of complicated surfaces to attain the excellent surface appearance and properties. In the last few decades, various advanced finishing processes have been developed for various applications that are summarized in Table 1.4. It shows the classification of advanced finishing techniques associated with their principle of finishing, advantages, disadvantages and potential uses (Jain, 2002; Jha and Jain, 2004; Rhoades, 1991; Jain, 2010; Fox et al., 1994).

AFM, abrasive flow machining; MAFM, magnetic abrasive flow machining; MRF, magnetic rheological finishing; MFP, magnetic float polishing; EEM, elastic emission machining; CMP, chemo-mechanical polishing; CMMRF, chemo-mechanical magneto-rheological finishing; MRAFF, magneto-rheological abrasive flow finishing; RMRAFF, rotational magneto-rheological abrasive flow finishing; ID, internal diameter; OD, outer diameter.

TABLE 1.4 Classification of advanced finishing techniques

PROCESS	PRINCIPLE	ADVANTAGES	DISADVANTAGES	APPLICATIONS
AFM	Micro-plowing and micro-cutting	Faster changeover media, fully automated, capable to finish both ID and OD	Higher initial cost, expensive fixture, no deterministic process	Aerospace, automobile, medical, precision dies, etc.
MAFM	Abrasion due to development of magnetic brush of iron particles	Easy cleaning, processed large parts, controlled media temperature	Difficult to finish flat surfaces, not processed blind holes	Optics, cutting tool, airfoils, turbine blades, sanitary pipes, needles etc.
MRF	Magnetic assisted mechanical abrasion using micron size carbonyl iron particles	Highly précised for finishing of lenses, Achieved up to 0.8 nm surface finish	Incapable to finish 3D shapes,	Optics, finishing of brittle materials, semiconductor wafers, etc.
MFP	Chemo-mechanical action (using magnetic fluid with abrasive)	Higher finishing rate, highly précised operation	Difficult to control, used for roller shape	Roller bearings, mechanical roller section, etc.
EEM	Chemical reaction shearing and erosion using elastodynamic fluid	Material removal up to atomic level, slurry used water as carrier, higher surface finish	Limitation in shape, wear of rotation sphere (polyurethane)	Finishing of aircraft parts, industrial applications, etc.

(Continued)

TABLE 1.4 (CONTINUED) Classification of advanced finishing techniques

PROCESS	PRINCIPLE	ADVANTAGES	DISADVANTAGES	APPLICATIONS
CMP	Chemical reaction and mechanical abrasion	Minimized surface damages, Increased IC reliability, Planarization	Time consuming, costly, required more attention at the time of operation, used only for flat surfaces	Semiconductor manufacturing, automobile industry and biotechnology, etc.
CMMRF	Chemical reaction and magnetic assisted mechanical abrasion (combination of CMP+MRF)	Achieved less than 1 nm finishing, high finishing rate, minimal surface defects	Not used to finish 3D surfaces, costly	Semiconductor industry to finish silicon blank, metrological probe, etc.
MRAFF	Magnetic assisted mechanical shearing and abrasion (combination of AFF+MRF)	Deterministic process, finish any complex geometry	Low finishing rate, non-uniform surface finish in free form surfaces	Optics, dies and mold, aircraft parts, etc.
RMRAFF	Combination of AFF+MRF+Rotation	Higher finishing rate, finish free form surface uniformly	Low finishing rate	Hip and knee joint, stents, etc.

1.9 NANOPARTICLES PRODUCTION TECHNIQUES

This hypothesis of nanotechnology was started in 1959 through the concept given by Nobel Prize laureate Richard Feynman. But it was first used and defined as nanotechnology by Japanese scientist Taniguchi in 1974. He stated that it mainly consists of the treating of separation, deformation and consolidation of materials by one atom or one molecule (Luther, 2004). In the 21st century, nanotechnology is the most promising technology to develop miniature products. After that scientists developed two routes to describe the various possibilities to produce nanostructures and nanoparticles (Darweesh, 2018). These routes are physical and chemical production routes that differ in their cost, speed and degree of quality. The various techniques of nanoparticles production are discussed in Chapter 3.

1.10 FUTURE OF ADVANCED MANUFACTURING

It is very important for the general public that they can understand the impact of advanced manufacturing in the society and economy of the country. It has played a significant role in deciding the gross domestic product (GDP) of a nation (Wang, 2018). In the future, industries are more concerned about the development of light metal for reduction of product weight without compromising its mechanical, thermal and tribological properties. So, more innovative materials require more modifications in current manufacturing processes with advanced meteorological tools. Further, the industries are more concerned about the sustainable aspect of manufacturing so that it does not harm our environment. Moreover, for higher production to meet the demand, the hybrid-type manufacturing can be more suitable in the near future. This can be achieved with an efficient integration between human and machine or robot in industry. Finally, the overall goal for the nation will be achieved through the proper collaboration between industry, academia and government.

REFERENCES

Ajay, C.V., Boopathi, C., and Kavin, P. (2019) Incremental sheet metal forming (ISMF): A literature review, *AIP Conference Proceedings* 2128, 030012.

Brickwood, R. (1995), Peen forming — a look under the surface, *Aircraft Engineering and Aerospace Technology*, 67 (4), 7–10.

Campbell Jr., F. (2003) *Manufacturing Processes for Advanced Composites*, Elsevier Science, Amsterdam.

Cao, J., Huang, Y., Reddy, N.V., Malhotra, R., and Wang, Y. (2008) Incremental sheet metal forming: Advances and challenges. *Paper presented at 9th International Conference on Technology of Plasticity*, ICTP 2008, Gyeongju, Republic of Korea, 16 p.

Charles, C., (1992) *Plaster Mold and Model Making*, Prentice Hall & IBD; Reissue edition.

Darweesh, H.H.M. (2018) *Nanomaterials: Classification and Properties-Part I*, Nano 1, 1.

Dorel, B. (2007) *Advanced Methods in Material Forming*, Springer Nature, Berlin, Heidelberg.

Fox, M., Agrawal, K., Shinmura, T., and Komanduri, R. (1994) Magnetic abrasive finishing of rollers, *Annals of CIRP*, 43(1), 181–184.

Gayakwad, D., Dargar M.K., Sharma P.K., Purohit R., and Rana R.S. (2014) A review on electromagnetic forming process, *Procedia Materials Science* 6, 520–527.

Ghosh, A. and Mallik, A.K. (2010) *Manufacturing Science*, East-West Press, Bangalore, India.

Giuliano, G. (2011) *Superplastic Forming of Advanced Metallic Materials*, Woodhead Publishing, Amsterdam.

Groover, M.P. (2018) *Principles of Modern Manufacturing SI Version*, Wiley India.

Hashmi S., Batalha G.F., Van Tyne C.J., and Yilbas B. (2014) *Comprehensive Materials Processing*, Elsevier.

Hingole, R.S. (2015) *Advances in Metal Forming*, Springer Nature, Berlin Heidelberg.

Jain, V.K., (2002) *Advanced Machining Processes*, Allied publishers, New Delhi (India).

Jain, V.K, (2010) *Introduction to Micromachining*, Narosa, India.

Jha, S. and Jain, V.K. (2004) Design and development of the magnetorheological abrasive flow finishing (MRAFF) process, *International Journal of Machine Tools and Manufacture* 44(10), 1019–1029.

Kalpakjian, S. (2014) *Manufacturing Engineering and Technology*, Fourth Edition, Addison-Wesley Publishing Co., Boston, MA.

Kirkwood, D.H., Suéry, M., Kapranos, P., Atkinson, H.V., and Young, K.P. (2010) *Semi-solid Processing of Alloys*, Springer Nature Berlin Heidelberg.

Lee, C.M., Kim, D.H., Baek, J.T., and Kim, E.J. (2016) Laser assisted milling device: A review, *International Journal of Precision Engineering and Manufacturing-Green Technology*, 3, 199–208.

Luther, W. (2004) *Industrial Application of Nanomaterials*, Future Technology Division, Germany.

Lysak, V.I. and Kuzmin, S.V. (2011) Lower boundary in metal explosive welding. Evolution of Idea, *Journal of Materials Processing Technology*, 212(1), 150–156.

Manikanda, P.K. and Vignesh, S. (2017) A review of advanced casting techniques. *Research Journal of Engineering and Technology*, 8(4), 440–446.

Mehta, K. (2017) *Advanced Joining and Welding Techniques: An Overview*, Springer Nature, Cham.

Metaxas, R.C. and Meredith, R.J. (1983) *Industrial Microwave Heating*, PeterPeregrinus Ltd., London.

Payal, H.S. and Sethi, B.L. (2003) Non-conventional machining processes as viable alternatives for production with specific reference to electrical discharge machining, *Journal of Scientific and Industrial Research* 62(7), 678–682.

Prasad, R.T.V. (2012) Progress in investment casting, In *Science and Technology of Casting Processes*, Srinivasan, M., Ed., Intechopen Publication, London, 28–38.

Rahim, E., Warap, N., and Mohid Z., (2015) *Thermal-Assisted Machining of Nickel-based Alloy*, Intech Open Publication, London.

Raymond, W.M. (1993) *Expendable Pattern Casting*, Amer Foundry Society, Schaumburg, IL.

Rhoades, L. (1991) Abrasive flow machining: A case study, *Journal of Materials Processing Technology*, 28(1–2), 107–116.

Sahu, M.K. and Sahu R.K. (2018) *Fabrication of Aluminum Matrix Composites by Stir Casting Technique and Stirring Process Parameters Optimization*, Intechopen Publication, London.

Soiński, M.S., Kordas, P., and Skurka, K. (2016) Trends in the production of castings in the world and in Poland in the XXI century, *Archives of Foundry Engineering*, 16(2), 5–10.

Sapanathan, T., Raoelison, R.N., Buiron N., and Rachik, M. (2016) *Magnetic Pulse Welding: An Innovative Joining Technology for Similar and Dissimilar Metal Pairs*, Intechopen Publication, London.

Singh, N., Mazumder, R., Gupta P., and Kumar, D. (2017) Ceramic matrix composites: Processing techniques and recent advancements, *Journal of Materials and Environmental Sciences*, 8(5), 1654–1660.

Viswanathan, S., Apelian, D., Donahue, R.J., Gupta, B.D., Gywn, M., Jorstad, J.L., Monroe, R.W., Sahoo, M., Prucha, T.E., and Twarog, D. (2008) *Casting, Volume 15, ASM Handbook*, ASM International, Geauga County, OH.

Wang, B. (2018) The future of manufacturing: A new perspective, *Engineering*, 4(5), 722–728.

Wise, R.J and Froment, I.D. (2001) Microwave welding of thermoplastic, *Journal of Material Science*, 36, 5935–5954.

Yuce, C., Karpat, F., Yavuz, N., and Dogan, O. (2015) A review on advanced joining techniques of multi material part manufacturing for automotive industry, *International Journal of Mechanical And Production Engineering*, 3(5), 63–68.

Zhang, F., Zhang, J., He, K., and Hang, Z. (2018) Application of electro-hydraulic forming (EHF) process with simple dies in sheet metal forming, *IEEE International Conference on Information and Automation (ICIA)*, Fujian.

Additive Manufacturing – "An Evolutionary Pace"

2

2.1 BACKGROUND OF ADDITIVE MANUFACTURING

Manufacturing has been known to be the foremost cause of economic growth of a nation. In recent times to become successful in the global manufacturing industry, innovative design and manufacture of products/parts/components in minimal time with least human intervention, minimum tooling, affordable cost, negligible wastage of material and acceptable quality has become a challenging task for industries. It is extremely difficult for the traditional manufacturing techniques like casting, forming, joining and machining to accomplish the above task. Therefore, industries shift their attention toward modern manufacturing techniques from traditional parts manufacturing methodology. Additive manufacturing (AM) has emerged as one of the modern manufacturing techniques and found to meet the aforementioned challenging task (Omar et al., 2015; Chua et al., 2005; Zhong 1999; Wiedemann and Jantzen, 1999; Dey and Yodo, 2019). AM is also known as rapid prototyping/automated fabrication/3DP/layered manufacturing/freeform fabrication techniques/other variants. AM technology has changed the landscape of industrial manufacturing. As compared to other modern manufacturing techniques, AM is a quicker and easier technique for innovative design and manufacture of intricate 3D models and prototypes with a reduction in labor and cost without

DOI: 10.1201/9781003203162-2

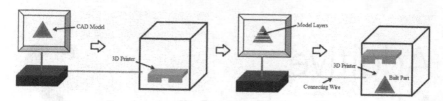

FIGURE 2.1 Fundamental steps used in AM of parts.

the intervention of tooling and fixturing. In contravention to traditional manufacturing, AM is independent of cost from the intricacy of parts. AM can often produce low-volume prototype functional parts by eliminating the time required for starting to fully operational stage of the manufacturing process. This technology enhances the manufacturing ability to change easily based on the situation and will use the maximum resources to manufacture more affordable parts.

AM can fabricate parts from the raw material in a solid/powder/liquid state. This technology is based on the additive layer mechanism, wherein the parts related to the virtual cross-section of its CAD model can be manufactured by adding the heated raw material layer by layer. The fundamental steps used in AM of parts (Figure 2.1) are as follows: (i) creation of CAD model of the part, (ii) conversion of CAD model into stereolithography (STL) file format, (iii) slicing the file into thin sections, (iv) building part layer by layer, and (v) joining/postprocessing/finishing of parts. In every marketable AM technique, the part is manufactured by deposition of layers contoured in a 2D X-Y plane whereas the third dimension (Z) effects from individual layers being piled up on top of each other, but as a discontinuous Z-coordinate (Onuh and Yusuf, 1999; Upcraft and Fletcher, 2003; Sood et al., 2009). The AM parts are found to be used in various applications such as aircrafts, space, automotives, medical, home appliances, electronics, computers, civil construction, textiles, sensors, biotech and displays.

2.2 GLIMPSE OF NON-METAL ADDITIVE MANUFACTURING TECHNIQUES

Over the past few decades, different non-metal-based AM techniques were developed and used by researchers to rapidly fabricate 3D physical objects (majority of them made from polymeric materials) of required size and shape from CAD data sources automatically without the application of tooling and least human intervention with reasonable time and cost. The various non-metal AM techniques include stereolithography, laminated object manufacturing (LOM), selective laser sintering (SLS), fused deposition modeling (FDM), solid ground curing (SGC), multi-jet modeling (MJM) and 3D ink-jet printing (Mansour and Hauge, 2003; Upcraft and Fletcher, 2003; Noorani, 2005; Sood et al., 2009; Sahu et al., 2013). The common important phases that are involved in the manufacturing of prototype parts using the non-metal AM techniques are shown in Figure 2.2. But the mechanism of the formulation of distinctive layers depends on a particular AM technique (Kai and Fai, 1997; Sood et al., 2009).

The beginning of any AM technique is the origin of the product geometry to be manufactured. The product geometry data set should then be collected and gathered and thus managed to produce the directives required to control the fabrication process of the part in the last phase. In general, any AM technique starts with creating a CAD file or solid modeling file (in STL file format) to represent the 3D shape of the product, i.e., STL file is produced by creating blocks of the 3D model. The created blocks are sliced into a multiple-layer data format and converting these layer-wise data into proper numerical codes that

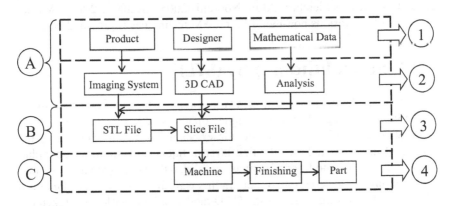

FIGURE 2.2 Common important phases applicable to almost all AM techniques.

1. Ideal product geometry origin.

2. Collection and gathering of geometrical data set.

3. Management of geometrical data.

4. Manufacturing of part.

A. Phases 1 and 2 do not depend on AM.

B. Phase consistent with almost all AM techniques.

C. Phase reliant on specific AM technique.

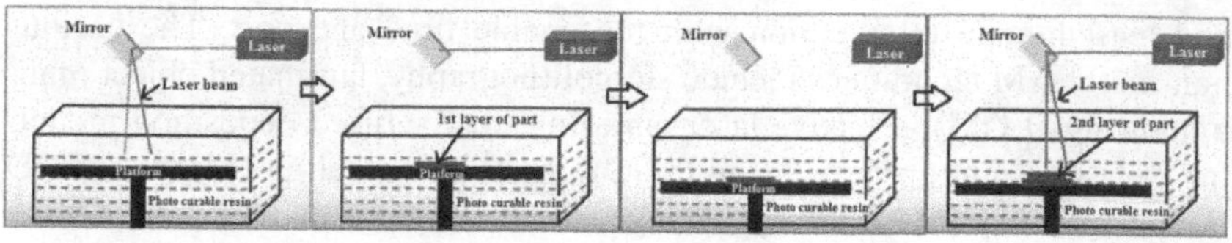

FIGURE 2.3   Fabrication of part using STL technique.

could be realized by AM machines. At this phase, since the time required for part building, surface quality, quantity of support structures, cost, etc. depend mainly on the parameters like the orientation of the part and slice thickness, the selection of these two parameters is of utmost importance. The numerical codes are then used to control the X–Y–Z movements of a material-depositing nozzle (e.g., in FDM technique) or a laser scanning head (e.g., in STL and SLS) and an object-supporting platform. Subsequently, corresponding to the virtual cross-section from the CAD model, each layer of the heated material is built on the preceding layer by each AM machine's particular material fabrication technology and is joined automatically to create the final model or part (Zhong et al., 2001; Sahu et al., 2013; Kai and Fai, 1997; Wang et al., 2000; Sood et al., 2009; Upcraft and Fletcher, 2003; Noorani, 2005; Tromans, 2003). A brief overview of the various non-metal-based AM techniques is provided below.

2.2.1 Stereolithography (STL)

Stereolithography (STL) is the first AM technique, which was launched in the USA two decades ago. In this technique, a photocurable liquid polymer resin as shown in Figure 2.3 is exposed to a laser beam. The beam scans across the surface of the resin, and thus the first layer of liquid resin solidifies on a platform. Then, the second layer of resin is exposed to the laser beam after the first solidified layer is lowered into a tank. In this way, the process is repeated until the solid 3D part is made as per the CAD model and the platform is taken out from the tank once the laser beam traces out the cross-section of all the slices. The completed part is cleaned off by the surplus liquid resin, and finally the part is then dried in an ultraviolet oven.

2.2.2 Selective Laser Sintering (SLS)

In selective laser sintering (SLS) technique, a thin layer of heat fusible powder (i.e., thermoplastic powder) is dispersed over the surface of a platform using a roller and heated to its solidus temperature by a laser beam

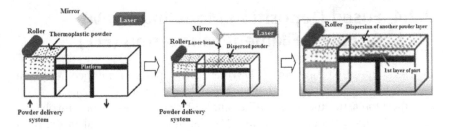

FIGURE 2.4 Fabrication of part using SLS technique.

(Figure 2.4). Then, the powder particles fused together on the platform, and the first fused layer is lowered into a cylinder followed by the dispersion of another layer of powder. In this way, the process is repeated until the solid 3D part is made by tracing out the cross-section of all the slices using the laser beam. The non-fused material in the individual layer can assist the part during fabrication, and when the part is completed, the non-fused material is cleaned off by brush.

2.2.3 Laminated Object Manufacturing (LOM)

In laminated object manufacturing (LOM) technique, one side of a sheet material layer is coated with an adhesive and the material with coated adhesive side is placed on a platform by a supply roller (Figure 2.5). A laminated roller (i.e., heated roller) passes over the material and adheres the material to the platform. The laser beam focuses on the material and cuts a layer of material as per the traced outline of a single slice of the part. Then, the platform with the first cut layer of material is moved down to a distance of first layer thickness and the second cut layer adheres onto the first layer. In this way, the process is repeated until all cross-section slices have been added, and finally the created solid 3D part is taken out from the platform.

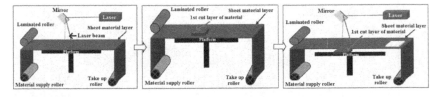

FIGURE 2.5 Creation of part using LOM technique.

2.2.4 Fused Deposition Modeling (FDM) or Fused Filament Fabrication (FFF)

In fused deposition modeling (FDM) technique, parts of any geometry can be built by sequential deposition of material on a layer-by-layer basis. FDM uses thermoplastic filaments unlike other AM techniques, which involve an array of lasers, powders and resins. The thermoplastic material in the form of a filament is supplied from a source spool and introduced into a channel of the nozzle. FDM uses two nozzles, one for part (model) material deposition and the other for support material deposition, and both works alternatively (Figure 2.6). The part material deposition nozzle is heated to melt and extrude out the thermoplastic material. The first layer of a solidifying material dispensed from the nozzle is deposited onto the surface of the platform (Figure 2.6) and upon completion of the first layer a second layer of material is then deposited onto the first layer and adhered thereto. Simultaneously, the support material deposition nozzle is used to build up a support structure for the part by extruding a different material. These procedures are repeated until the 3D solid part is created, and finally the support material is broken away from the created part.

2.2.5 Solid Ground Curing (SGC)

In solid ground curing (SGC) technique, a mask is prepared on a glass substrate as per the CAD model data. The glass is placed over the photopolymer

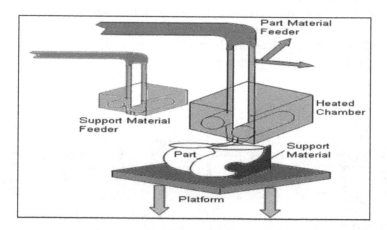

FIGURE 2.6   Schematic diagram of FDM technique used for part creation.

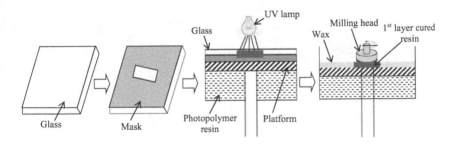

FIGURE 2.7   Fabrication of part using SGC technique.

resin and exposed to ultraviolet light (Figure 2.7). Then, a layer of resin is cured on the platform and the uncured resin is removed. The wax is applied on the empty spaces remained after the uncured resin is removed away. The surface is then milled to obtain the required surface finish and thickness. The cooled wax chips are removed through milling. A new layer of photopolymer resin is again exposed to UV light, and the cured layer adheres to the previous layer (Figure 2.7). In this way, the process is repeated until the final prototype is made. In this technique, the time required for the formation of multiple prototypes is less and no post-curing is required, but it is noisy and a large amount of wax is wasted.

2.2.6 3D Ink-Jet Printing

The 3D ink-jet printing is a technique, which builds parts in layers using ink-jet technology. In this technique, a thin layer of powdered material is dispersed uniformly over the surface of a powder bed using a roller (Figure 2.8). The ink-jet impinges on a particular area of the layer as per the traced cross-section of a single slice and the binder material joins the powder particles. Then, the bed with the first layer of formed material is moved down to a fixed distance and the second layer is built onto the first layer. In this way, the process is repeated until the solid 3D part is formed. The final part is taken out by brushing the unprocessed powders wherein the part is enclosed.

2.2.7 Multi-Jet Modeling (MJM)

In multi-jet modeling (MJM) technique, a series of nozzles (96 Nos.) are present in a print head and each nozzle distributes the heated thermoplastic polymer in the form of a droplet at the same time over a platform. Thus, the first layer of material is formed on a platform as per the cross-section of a

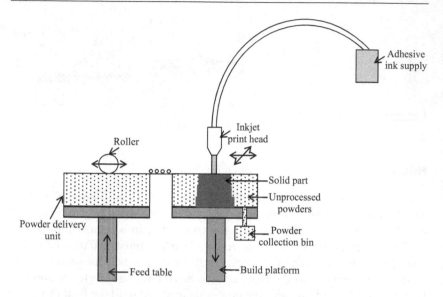

FIGURE 2.8 Fabrication of part using 3D ink-jet printing technique.

single slice. Then, the platform with the first layer of formed material is moved down relative to the print head and the second layer is bonded to the first layer. Simultaneously, the support material deposition nozzle is used to build up a support structure for the part by extruding a different material. These procedures are repeated until the 3D solid part is created, and finally the support material is broken away from the formed part. In this technique, the rate of dispense of hot material droplets over a platform was found to be fast.

The non-metal-based AM techniques as described above are usually identical to each other that can produce the objects by adding and bonding materials in a layer-wise manner. This mechanism is contrary to that of traditional material removal and forming techniques. The process-related parameters used in various AM techniques to fabricate polymer-based components/parts include fixed parameters like part fill style, contour width, part interior style, visible surface, part xyz shrinkage factor and perimeter to raster air gap and control parameters like layer thickness, part orientation, raster angle, raster width, raster to raster gap (air gap), infill rate and the number of shells. The various polymeric materials used in AM techniques are acrylonitrile butadiene styrene (ABS), polyether ether ketone (PEEK), polylactic acid (PLA), polydimethylsiloxane (PDMS), polyethylene terephthalate (PETG), polyvinyl alcohol (PVA), nylon, flex, some high-temperature polymers, etc.

The large-scale industries have used the above AM techniques in remarkable ways to address different problems in various technical fields (Chua et al., 1999; Liu et al., 2005; Raja et al., 2005; Sood et al., 2009). Besides,

TABLE 2.1   Pros and cons of AM

SL. NO.	PROS	CONS
01	Design freedom	Degraded dimensional control and surface integrity
02	Production of complex geometries and internal features	Low standardization
03	Reduction of time to market	Produced parts often require postprocessing
04	Toward mass customization	Limited materials are available for processing
05	Elimination of tooling	-
0G	Reduction of material wastage and energy consumption	-
07	Reduction in part counts, handling time and storage requirement and avoids mating and fit problem	-

small- and medium-scale industries were not able to use the techniques in their shop floor, because it became hard for them to procure the high-expensive AM machines. But, with the advent of internet-based AM technologies, these industries can able to manufacture components/parts remotely without procuring the AM machines (Lan, 2009). Now, owing to the wide usage of the internet, it becomes a standard that the prospect of manufacturing industries will be directed toward information, knowledge driven and most of their regular processes will be automated across the world information network that joins each one together (Dong et al., 2008). The common pros and cons of the non-metal-based AM techniques are shown in Table 2.1.

2.3 ADVENT OF METAL ADDITIVE MANUFACTURING TECHNIQUES

The non-metal-based AM techniques have been used for fabrication of non-metallic, in particular, polymer-based parts/components of desired shape and size, and are used in various multidisciplinary emerging applications. However, innovative design and manufacture of products made of metallic and metallic alloy-based advanced material products with acceptable quality, at affordable costs and in minimal time has currently gaining a remarkable research and market attention. These requirements are sweeping the world of manufacturing. It was found that the non-metal-based AM techniques are not capable to manufacture these materials. Thus, the advent of metal-based AM technologies is offering manufacturers new paths to design and manufacture an innovative range of metal-based consumer products, save time, conserve energy, expand human horizons, and compete across the global industrial market.

Therefore, metal AM (metal 3DP) has become a hot topic among researchers and engineers in recent years. Metal 3DP presents a cutting-edge technology to manufacture metallic components via additive fashion. To date, the technology was not so developed to exploit widely in the area of manufacturing. Nevertheless, due to the recent advancement in technology, the metal 3DP is changing into a measurable and firm solution. The metal 3DP can offer various technical and business benefits like the design of geometrically complex parts, manufacture metallic and metallic alloy-based parts without tooling, produce parts without detail drawings, print customized features in an affordable manner, get the products to market faster, reduce the manufacturing costs and replace inefficient manufacturing workflows. This technology almost does not require any supervision while fabricating parts. Further, metal 3DP can be able to produce high-performing components from difficult-to-difficult machine materials like superalloys at low costs in comparison with subtractive methods.

2.4 CONCEPTUAL REALIZATION OF METAL ADDITIVE MANUFACTURING TECHNIQUES

The metal AM or metal 3DP technology usually uses metallic powder or metal powder as media and changes the powder into solid metal components using energy sources ranging from high-energy laser pulses to extruding bound metal powder filament/fused bound metal powder deposition (FBMPD). The metal 3DP techniques that are recently launched in the market to produce metallic-based products include powder bed fusion (PBF), direct energy deposition (DED), binder jetting, bound powder extrusion (BPE) (bound powder deposition), and wire arc additive manufacturing (WAAM) (Shirizly and Dolev, 2019; Tomer et al., 2020; Rodrigues et al., 2019a; Rodrigues et al.,

2019b; Zakay and Aghion, 2019; Leon and Aghion, 2017; Gu, 2015; Gordon et al., 2019). Each type of these techniques is a unique one and has its own advantages and limitations.

The common procedure involved in the fabrication of metal-based components using different metal AM (metal 3DP) techniques is just like polymer-based 3DP, wherein all metal 3D-printed parts start with an STL file. The STL file can be exported from CAD software or created with a 3D scanner and uploaded to cloud slicing software. After uploading, the part can be oriented by clicking on a face to mate it to the build platform/build table. Once the part orientation is selected and the metal is chosen, cloud slicing software slices the part. Then, the metal part is built layer by layer followed by postprocessing of the built part. During the printing process where the part often needed any support, the cloud slicing software generates support material in metal below them. Also, recently researchers used a ceramic material release layer (generated using cloud slicing software) between the part and support that become powder after sintering and can be helpful to remove the part easily with the light tap of a hammer after sintering. Moreover, in order to avoid the bend of the part during printing and to have proper shrink during sintering, a raft (generated using cloud slicing software) is used below both the part and the support. The conceptual realization of various metal AM or metal 3DP techniques is briefly discussed below.

2.4.1 Powder Bed Fusion Technique

Powder bed fusion (powder bed melting) technique is the most commonly used metal 3DP. There are two distinct types of PBF techniques – Direct Metal Laser Sintering (DMLS) and Electron Beam Melting (EBM). The DMLS is also known as Direct Metal Printing (DMP) or Laser Powder Bed Fusion (LPBF). In PBF technique, a fine layer of loose metal powder (e.g., steel and bronze) is distributed over a powder bed (build plate). Then, a high-power laser source or electron beam is used to selectively melt loose metal powder and fuse metal layers into parts (Figure 2.9). Finally, the part is removed from the powder bed and postprocessed to the required shape and size.

The produced parts from this technique fit into a wide variety of applications ranging from dental/healthcare to aerospace. In PBF, the maturity level is high, part size – 2–350 mm and precision is high. The precision of DMLS and EBM machines is determined by laser beam width and electron beam width, respectively, and also the layer height. Today, most of the metallic materials available to be 3D printed can be used on DMLS machines. The major difference between DMLS and EBM techniques is that the electron beam yields a less precise part than DMLS, but the EBM technique as a whole is faster for

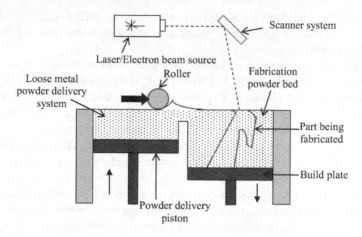

FIGURE 2.9   Fabrication of part using PBF technique.

larger parts. The common advantages and disadvantages of PBF techniques are as follows:

Advantages
  i. A variety of selection of machines and materials.
 ii. Able to produce high-quality intricate parts.
iii. Repeatability is high.

Disadvantages
  i. Setup cost is high, and labor-intensive is required to function.
 ii. Postprocessing treatment of parts is required, and warping/cracking usually occurs.
iii. Parts are welded to the build platform and must be removed with electrical discharge machining, and also support material must be cut off.
 iv. Changing of loose metal powders requires hours of labor.

2.4.2   Direct Energy Deposition Technique

In direct energy deposition (DED) technique, metallic powder or wire and a laser is used to build parts. There are two distinct types of DED techniques – Powder DED and Wire DED. Unlike PBF technique, in DED powder/wire and laser both sit on a single print head. Instead of spreading powder on a bed and melting it with a laser, in powder DED the metal powder is blown out

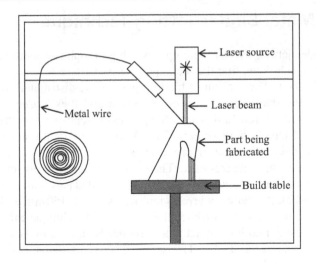

FIGURE 2.10   Fabrication of part using DED technique.

of a print head onto a build table and at the same time the material is fused by a laser layer-by-layer to form a part. But, in the case of wire DED, the metal wire is fed through a print head and the laser fuses the wire simultaneously to form a final part (Figure 2.10). Finally, the part is removed from the build table and postprocessed to the required shape and size. In DED, the maturity level is high, part size – 50–2000 mm and precision is low. Powder DED technique is slightly faster and less accurate than PBF technique whereas wire DED technique is not similar to PBF, but wire DED can be able to produce very large rough shaped parts faster at the expense of precision and quality as compared to powder DED. Further, powder DED technique is somewhat comfortable than PBF because the former can utilize its unique powder distribution system to "heal" non-printed parts that have deficiencies. The common advantages and disadvantages of DED techniques are as follows:

Advantages
  i. DED is faster than PBF technique because of the large nozzles and as a result the flow rate of the powder is fast.
  ii. In DED technique, the material can be added dynamically on different planes.
  iii. It can repair the broken components dynamically.

Disadvantages
  i. Setup cost is high, and its function is labor-intensive.
  ii. DED is less precise and accurate than PBF technique.

2.4.3 Metal Binder Jetting Technique

Metal binder jetting (MBJ) technique is a high-reliability method of metal 3DP. In this technique, first the metal powder particles are evenly distributed over a print bed. Then, a liquid binding polymer is distributed through the jet head in the shape of the part cross-section and thus loosely adhered to the metal powder layer. In this way, the process is repeated until the finished light-bounded part is built (Figure 2.11). Further, the light-bounded printed part is subjected to the sintering process where the part is heated in an oven to just below its melting temperature. The binding material burns away and the metal powder unite into a full metal part. The bounded parts can be sintered in batches. In MBJ, the maturity level is low, part size – 1–150 mm and precision is high. As compared to the PBF technique, the MBJ technique can print parts much faster using multiple ink-jet type print heads and can make hundreds of the same part in one build table. The advantages and disadvantages of the MBJ technique are as follows:

Advantages
   i. Less energy is required because of the absence of lasers.
  ii. Quickly bind the layers together using an ink-jet-type print head.
 iii. High production rate as compared to other techniques.
  iv. Extremely complicated geometries can be created without any requirement of mechanical separation from the build platform.
   v. It is cost-effective than PBF.

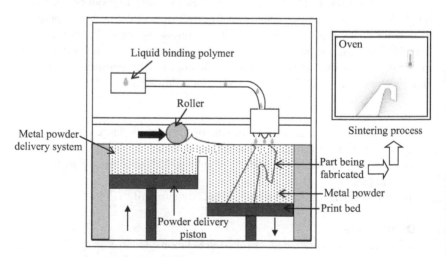

FIGURE 2.11   Fabrication of part using MBJ technique.

Disadvantages
  i. Accuracy, quality and repeatability are yet to be explored in detail.
  ii. Prior to sintering, the light-bounded printed parts are very delicate.

2.4.4 Bound Powder Extrusion (Bound Powder Deposition) Technique

Bound powder extrusion (bound powder deposition) technique is also known as FBMPD technique or atomic diffusion additive manufacturing (ADAM) technique. BPE or ADAM is an exciting newly introducing technique to the metal AM family. BPE technique closely related to the combination of metal injection molding and fused filament technologies. Unlike almost every other metal 3DP technique, BPE technique does not use loose metal powder. This technique uses metal powder where powder atoms are bound together in waxy polymers (i.e., bound powder filament) in a similar manner as metal injection molding stock.

The bound powder filament (also called BPE filament) is extruded out of a nozzle and deposited layer by layer on the build table. Thus, a green part is formed that contains metal powder uniformly distributed in waxy polymer. Then, the green part is placed in a washing system where the polymer binder is dissolved and the washed part (brown part) is sintered in an oven (Figure 2.12). During the sintering process, the part shrinks and finally formed into a fully metallic part. In BPE technique, the maturity

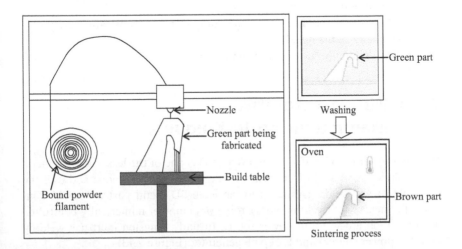

FIGURE 2.12 Fabrication of part using BPE technique.

level is medium, part size – 10–250 mm and precision is medium. The BPE/ADAM technique is capable of building fully enclosed lattice/mesh structures, resulting in parts with high strength-to-weight ratios, but in the case of PBF technique unprocessed material can be trapped in such enclosed mesh structures, making it difficult or impossible to remove. The BPE technique can be suitably used in various manufacturing applications owing to its low cost and ease of use. Further, in BPE, recently industrial researchers are using a ceramic release layer (in the form of powders) in between the green part and support material and also a raft below the support and green part. This ceramic release layer can be useful to remove the green part easily from the support material without any distortion to the part. The common control parameters used in this technique and also in PBF, DED and MBJ techniques are almost the same as used in non-metal (in particular, polymer) based AM techniques.

Advantages
  i. Cost-effective and considerable ease of operation.
  ii. High-yield parts can be produced at the first instance of print.
  iii. Bound powder filament is safe and easy to use than loose powder.
  iv. Able to print a wide variety of pure metallic parts, which are hardly obtained by other techniques.
  v. It is cost-effective than PBF.
  vi. Near net shaped parts could be produced.

Disadvantages
  i. Low production rate.
  ii. The size and complexity of the component are constrained by the diameter of the nozzle.
  iii. The debinding time of large parts is more.

2.4.5 Wire Arc Additive Manufacturing Technique

Wire Arc Additive Manufacturing (WAAM) is one of the lesser-known metal 3DP techniques. WAAM is a variation of a wire DED technique and uses the principle of arc welding process to fabricate 3D metal parts. WAAM consists of a metal wire (welding wire), robot or computer numerically controlled machine tool, arc welding power source (usually tungsten inert gas welding (TIGW) power source) and tool path generator (Figure 2.13).

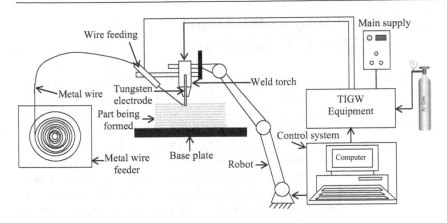

FIGURE 2.13 Fabrication of part using WAAM technique

Unlike the other metal AM techniques, WAAM works by melting metal wire using an electric arc as the heat source under an open or inert gas atmosphere. The melted wire is extruded in the form of beads on a substrate material (base plate). As the beads stick together, they create a layer of metal material. Then, the process is repeated by layer-by-layer material deposition carried out by a robotic arm with a programmed pathway controlled by a CAD model until the metal 3D part is completed (Figure 2.13). Finally, this metal 3D preform is followed by postprocessing (i.e., heat treatment and machining) of the part. The process-related parameters in WAAM include layer thickness of slice, deposition speed, voltage, current, pulse frequency wire feed rate and protective gas atmosphere.

Advantages

    i. Ability to print large 3D metal parts.

    ii. Cost-effective technique and the metal wires used are significantly less expensive than the metal powders used in PBF technique.

    iii. High deposition rate of raw material.

    iv. Produce dense and strong mechanical properties parts with negligible porosity.

    v. Worn-out features or damaged parts can be repaired by depositing new material on its surface.

Disadvantages

    i. Residual stresses are induced that often lead to distortion of metal parts.

    ii. Gas shielding is required for some materials (e.g., titanium).

    iii. Produce near-net-shape parts with poor surface finish.

2.4.6 Promising Applications of Metal 3D Printing

The materials that are used to produce metal 3D-printed parts include 17-4 PH stainless steel, 303 stainless steel, maraging steel, H13 tool steel, A2 and D2 tool steel (stainless steel, maraging steel and tool steel are more costly and difficult to fabricate traditionally), superalloys like Inconel 625, cobalt chrome, Ti–6Al–4V titanium alloy, titanium 316L, pure copper and its alloys, aluminum and its alloys 6061 and 7075 and magnesium alloys. The metal and its alloy-based 3D-printed parts/components fabricated using the above-mentioned metal-based AM techniques are widely used in various promising applications. Indicatively, such applications include but are not limited to

- Aerospace components (e.g., turbine blades)
- Marine (propeller, wear rings, etc.)
- Automotives (e.g., metal wheel-set bearing cover)
- Consumer package goods
- Electronics
- Defense
- Punch and dies
- PG10 profile grinder to grind railway tracks
- Impeller
- Robot end effectors (metal grippers)
- Surgical tools
- Orthopedic implant removal tool and other medical devices
- Crucible clips
- Special cutting tools
- Post driver
- Jigs and fixtures
- Seamless flow-formed cylinders (closed cone and parabolic cylinders)
- Crane hook
- Bike frame
- Household appliances
- Energy
- Nuclear industries
- Oil and gas
- Industrial equipment
- E-motor cycle, etc.

2.5 FUTURE DIRECTIONS

Metal 3DP can offer benefits to industrial manufacturers like overcoming increased overhead costs and producing custom tools at a low cost per piece that could be available in a short span of days. Although metal AM (metal 3DP) techniques currently offer promising opportunities, there are still a number of technical issues on the road to commercialization. Some technical limitations/disadvantages of commercially available metal 3DP machines were identified in Sections 2.4.1–2.4.5, where some research needs to be carried out to overcome these limitations and to accomplish efficient performance objectives. There is a need to explore new materials (e.g., titanium 316L, which is in the development stage in metal AM) in addition to developing hybrid metal wire or hybrid bound powder filament materials, which would be another good direction of research.

However, many industries have a strong need for large volume production of metal 3DP products in an economical and efficient way. Therefore, it is necessary to focus on ADAM and WAAM techniques, which are lesser-known techniques to date, and to explore the techniques that could meet the stringent demand. These newly proposed techniques could have the potential to enable the use of different materials within their process session. Deep understanding and good tuning of the techniques could yield metal 3D parts with superior performance characteristics.

Metal 3DP could be coupled with fiber 3DP to create hybrid techniques, which could leverage the benefits of both techniques to create higher-performing techniques. Metal 3DP could be integrated with a machining system so that the parts created by 3DP can be immediately finished with machining. There is a need for research on a closed-loop feedback system through image processing and machine learning algorithms, which could be able to control and monitor metal 3DP process performance. The realization on the integration of other advanced disciplines with the metal AM techniques and their application in an Industry 4.0 will be a viable research in the near future. Further, the development of AM techniques at the nanoscale will provide a future path for new nanotechnology applications. Moreover, the diversity in today's market widens as researchers are gaining knowledge of how to modify the materials and manufacture them, in order to take the advantage of various commercial applications.

These future directions will open the window for new research and developments required to establish the full scope of metal AM techniques approach for various engineering applications with high part quality, accuracy and desired properties.

2.6 SUMMARY

In this chapter, the background of AM and a glimpse of the non-metal-based, in specific, polymeric material-based AM techniques were briefly discussed. The advent of metal AM (metal 3DP) techniques followed by the conceptual realization of its various techniques currently developed to manufacture metallic-based components was discussed. Further, the promising applications of metal 3DP are outlined and the future directions are presented.

REFERENCES

Chua, C.K., Feng, C., Lee, C.W., and Ang, G.Q. (2005) Rapid investment casting: Direct and indirect approaches via model maker II, *International Journal of Advanced Manufacturing Technology*, 25, 11–25.

Chua, C.K., The, S.H., and Gay, R.K.L. (1999) Rapid prototyping and virtual manufacturing in product design and manufacturing, *The International Journal of Advanced Manufacturing Technology*, 15, 597–603.

Dey, A. and Yodo, N. (2019) A systematic survey of FDM process parameter optimization and their influence on part characteristics, *Journal of Manufacturing and Materials Processing*, 3, 64, 1–30.

Dong, B., Qi, G., Gu, X., and Wei, X., (2008) Web service oriented manufacturing resource applications for networked product development, *Advanced Engineering Informatics*, 22, 282–295.

Gordon, J., Hochhalter, J., Haden, C., and Harlow, D.G. (2019) Enhancement in fatigue performance of metastable austenitic stainless steel through directed energy deposition additive manufacturing. *Materials and Design*, 168, 107630.

Gu, D. (2015) Laser additive manufacturing (AM): Classification, processing philosophy, and metallurgical mechanisms. In *Laser Additive Manufacturing of High-Performance Materials*; Springer: Heidelberg/Berlin, Germany, 15–71.

Kai, C.C. and Fai, L. K. (1997) *Rapid Prototyping: Principles and Applications in Manufacturing*, John Wiley and Sons, Singapore.

Lan, H. (2009) Web based rapid prototyping and manufacturing systems: A review, *Computers in Industry*, 60, 643–656.

Leon, A. and Aghion, E. (2017) Effect of surface roughness on corrosion fatigue performance of AlSi10Mg alloy produced by selective laser melting (SLM). *Materials Characterization* 131, 188–194.

Liu, Q., Leu, M.C., and Schmitt, S.M. (2005) Rapid prototyping in dentistry: Technology and application, *The International Journal of Advanced Manufacturing Technology*, 29, 317–335.

Mansour, S. and Hauge, R. (2003) Impact of rapid manufacturing on design for manufacturing for injection moulding, *Proceedings of the Institution of Mechanical Engineers: Journal of Engineering Manufacture, Part B*, 217(4), 453–461.

Noorani, R. (2005) *Rapid Prototyping-Principles and Application*, John Wiley & Sons, New Jersey, USA.

Omar, A.M., Syed, H.M., Jahar, L.B. (2015) Optimization of fused deposition modeling process parameters: a review of current research and future prospects, *Advances in Manufacturing*, 3, 42–53.

Onuh, S.O. and Yusuf, Y.Y. (1999) Rapid prototyping technology: Applications and benefits for rapid product development, Journal of Intelligent Manufacturing, 10, 301–311.

Raja, V., Zhang, S., Garside, J., Ryall, C., and Wimpenny, D. (2005) Rapid and cost-effective manufacturing of high integrity aerospace components, The International Journal of Advanced Manufacturing Technology, 27, 759–773.

Rodrigues, T.A., Duarte, V., Avila, J.A., Santos, T.G., Miranda, R.M., and Oliveira, J.P. (2019a) Wire and arc additive manufacturing of HSLA steel: Effect of thermal cycles on microstructure and mechanical properties, *Additive Manufacturing* 27, 440–450.

Rodrigues, T.A., Duarte, V., Miranda, R.M., Santos, T.G., and Oliveira, J.P. (2019b) Current status and perspectives on wire and arc additive manufacturing (WAAM), *Materials* 12, 1121.

Sahu, R.K., Mahapatra, S.S., and Sood, A.K. (2013) A study on dimensional accuracy of fused deposition modeling (FDM) processed parts using fuzzy logic, *Journal of Manufacturing Science and Production*, 13(3), 183–197.

Shirizly, A. and Dolev, O. (2019) From wire to seamless flow-formed tube: leveraging the combination of wire arc additive manufacturing and metal forming, *JOM*, 71(2), 709–717.

Sood, A.K., Ohdar, R.K. and Mahapatra, S.S. (2009), Improving dimensional accuracy of fused deposition modeling processed part using grey Taguchi method, *Materials and Design*, 30, 4243–4252.

Tromans, G. (2003) *Developments in Rapid Casting*, Professional Engineering Publishing, London, UK, 2003.

Tomer, R., Levy, G.K., Dolev, O., Leon, A., Shirizly, A., and Aghion, E. (2020) The effect of microstructural imperfections on corrosion fatigue of additively manufactured ER70S-6 alloy produced by wire arc deposition, *Metals*, 10, 98, 1–10.

Upcraft, S. and Fletcher, R. (2003) The rapid prototyping technologies, *Assembly Automation*, 23(4), 318–330.

Wang, W.L., Conley, J.G., Yan, Y.N. and Fuh, J.Y.H. (2000) Towards intelligent setting of process parameters for layered manufacturing, *Journal of Intelligent Manufacturing*, 11, 65–74.

Wiedemann, B. and Jantzen, H.A. (1999), Strategies and applications for rapid product and process development in Daimler-Benz AG, *Computers in Industries*, 39(1), 11–25.

Zakay, A. and Aghion, E. (2019) Effect of post-heat treatment on the corrosion behavior of AlSi10Mg alloy produced by additive manufacturing. *JOM* 71, 1150–1157.

Zhong, W.H. (1999), Rapid prototyping manufacturing technology and its development, *Aerospace Materials and Technology*, 29(3), 23–26.

Zhong, W.H., Li, F., Zhang, Z.G., Song, L., and Li, Z. (2001), Short fiber reinforced composites for fused deposition modeling, *Journal of Materials Science and Engineering A*, 301, 125–130.

Advanced Machining in the Age of Nanotechnology

3

3.1 NANOTECHNOLOGY IN CURRENT MILLENNIUM

In this millennium, nanotechnology has grown by leaps and bounds, and is regarded as one of the most multidisciplinary research and development fields undergoing explosive development worldwide. Nanotechnology deals with materials or structures at the ultrafine level. These materials/structures will provide more functionality in a given space. However, different people have different opinions about nanotechnology. Some people say the study of the microstructure of materials using a microscope is considered nanotechnology. Some other people say the study of the growth of thin films is defined as nanotechnology. Some other people say the drug delivery systems, MEMS/NEMS (Micro-electro Mechanical System/Nano-electro Mechanical System) devices and lab on a chip are considered nanotechnology. In the USA, people say the materials and systems whose structures and components are in nanometer size are considered nanotechnology. But, in general, nanotechnology can be defined as the technology for production, characterization and applications of nanomaterials. This indicates that, in particular, it has the potential

DOI: 10.1201/9781003203162-3

for transforming the materials at the nano-level scale known as nanomaterials (Darweesh, 2018; Sahu and Hiremath, 2017; Jalill et al., 2016). Nanomaterials are those materials that are having at least one dimension in the nanometer scale, i.e., less than 100 nm. This description of nanomaterials is probably the most intuitive one and cannot be exhaustive, as it does not present any specific values (Strambeanu et al., 2015; Jalill et al., 2016). Various research organizations have expressed the meaning of nanomaterials in a different fashion (Jalill et al., 2016).

A wide variety of nanomaterials are available that are used in various emerging applications. These materials include nanoparticles, nanorods, nanotubes, nanowires, nanofibers, thin films, graphene and nanocomposites. Table 3.1 shows the classification of nanomaterials based on their dimension and phase structure (Luther, 2004). The classification of nanomaterials based on their origin is shown in Figure 3.1 (Al-Kayiem et al., 2013). Among the various types of nanomaterials, the field of nanoparticle technology has attracted a lot of research attention in recent times and has become the fast-growing field because of the promising applications of nanoparticles.

TABLE 3.1   Classification of nanomaterials based on their dimension and phase structure

CLASSIFICATION	EXAMPLES
Based on Dimension	
0D nanomaterials	Nanoparticles – spheres, cubes, polyhedral, etc., quantum dots, graphene, fullerenes, etc.
1D nanomaterials	Nanotubes, nanorods, nanowires, nanofibers, etc.
2D nanomaterials	Graphene, nanofilms, nanolayers, nanocoatings, etc.
3D nanomaterials	Bulk powders, bundles of nanotubes and nanowires, multilayers, etc.
Based on Phase Structure	
Single-phase solid nanomaterial	Crystalline, amorphous nanomaterials and layers
Multi-phase solid nanomaterial	Core–shell nanomaterials and functionalized nanomaterials
Multi-phase system nanomaterial	Colloids, aerogels, aerosols, etc.

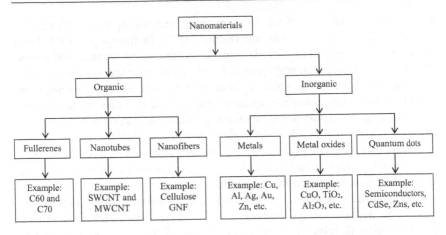

FIGURE 3.1   Classification of nanomaterials based on their origin.

3.2 SCIENCE OF NANOPARTICLES

Nanoparticles have gained considerable importance in current years owing to their exceptional physical, thermal, mechanical, electrical, optical, magnetic properties, catalytic, biomedical properties, etc. and thus potentially used in industrial, scientific and medical applications. Nanoparticles are nothing but they are microscopic materials whose dimensions usually lie in the range of 1–100 nm. These particles bridge the gap between bulk materials and molecules. Table 3.2 shows the correlation between the size range of molecules,

TABLE 3.2   Correlation between size range of molecules, nanoparticles and bulk of the same material

MATERIALS	NUMBER OF ATOMS	SIZE (NM)
Molecules	1	0.1
	10	
Nanoparticles	100	1
	1000	100
	10000	
	1 lakh	
	1 million	
Bulk	>1 million	>100

nanoparticles and bulk of the same material and the approximate number of atoms they contain (Poole and Owens, 2003). In the range of 0.1–1 nm, molecules contain a cluster of 10 atoms or more, but fewer than 100 atoms. Nanoparticles contain an aggregate of 1 million atoms or less in between 1 and 100 nm, which can be viewed as a subdivision of bulk materials having more than 1 million atoms. However, quantum mechanical Heisenberg's uncertainty principle as applied only to the subatomic particles places no limit on how well atoms and molecules can be held in place (Mansoori, 2005).

Nanoparticles present a practical way of retaining the results of the property at the atomic or molecular level. It shows a great promise for providing us in the near future with many breakthroughs that will change the direction of technological advances in a wide range of applications. Nanoparticles play a noteworthy role in the present century due to their increased size-dependent characteristics as compared to their bulk form of the same material. Their exceptionality occurs from their large specific surface area showing that a high percentage of atoms present on the particles surface over its whole volume (Gleiter, 2000; Sahu and Hiremath, 2017). This key feature calls for realizing the increased free surface energy of nanoparticles when scaled down from the bulk material.

As it has been known that a bulk material consists of atoms/molecules. When a bulk material is broken into fine particles, the fine particles tend to be affected by the behavior of atoms/molecules themselves and show different properties as compared to their bulk form. It is attributable to the change in the bonding state of the atoms/molecules constituting the particles. In other words, this indicates that the atoms at the surface of a nanoparticle (i.e., outermost atoms) are in a very different environment as compared to those atoms from its interior. This difference arises from the asymmetrical environment. But, in the bulk material, each atom is surrounded by similar ones and they experience no net forces. In addition, the various influencing factors exerted by the environment act only on the outermost atoms. As a result, the outermost atoms exhibit different energy distribution than the interior atoms and are in a high energy state at the surface. The difference between the energies of atoms located at the particles surface and energies of atoms inside the solid particles is known as free surface energy. The free surface energy is the excess energy at the surface of a nanoparticle as compared to those inside the particle and accordingly the particle can behave extraordinarily in relation to its bulk form. It is also observed that when the size of the particles approaches the nanoscale with characteristic scale length close to or smaller than the de Broglie wavelength of the charge carriers or wavelength of light, the periodic boundary conditions of the crystalline particles are destroyed or the surface atomic density of amorphous particles is changed. These reasons are responsible for different physicochemical properties than the bulk material, leading to various applications (Guo et al., 2013; Sant, 2012). Figure 3.2 shows the fundamental constituents of a nanoparticle (Heinz et al., 2017).

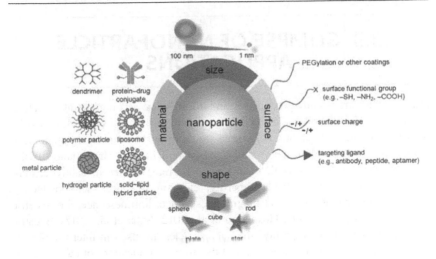

FIGURE 3.2 Fundamental constituents of a nanoparticle.

Generally, the materials such as metals, non-metals, semiconductors, polymers and ceramics can be modulated into nanoparticles. A set of different forms of nanoparticles ranges from inorganic to organic, crystalline to amorphous particles, aggregates, powders or dispersion in a matrix, suspensions and emulsions, class of fullerenes and their derivatives. The nanoparticles fall under the category of 0D nanomaterials, wherein all the dimensions are in the nanometer range. Many researchers have used different approaches to classify nanoparticles and usually the nanoparticles are classified based on their dimensional shape, metal and metal oxide-based particles and nature of bonding between the atoms of the particles (Reghunadhan et al., 2012; Maier, 1990; Santos et al., 2015; Johnston and Wilcoxon, 2012; Sahu and Somashekhar, 2019).

Based on the dimensional shape, the nanoparticles are typically considered 2D-shaped nanoparticles (e.g., circular, triangular and polyhedron) and 3D-shaped nanoparticles (e.g., spherical, oval, prismatic, cubic and helical/pillar). The metal-based nanoparticles include gold (Au), silver (Ag), platinum (Pt), palladium (Pd) alloy, copper (Cu), iron (Fe), nickel (Ni), cobalt (Co), aluminum (Al), zinc (Zn), manganese (Mn), molybdenum (Mo), tungsten (W), lanthanum (La), lithium (Li) and rhodium (Rh). The metal oxide-based nanoparticles include CuO, ZnO, TiO_2, SnO_2, Al_2O_3, MgO, AgO, CeO_2 and ZrO_2. Based on the nature of bonding between the atoms, the nanoparticles are classified as covalently bonded (e.g., semiconductor nanoparticles), van der Waals' bonded (e.g., molecular nanoparticles), ionic bonded (e.g., ionic nanoparticles), metallic bonded (e.g., metal nanoparticles and nano alloys) and coordinate bonded (e.g., capped nanoparticles).

3.3 GLIMPSE OF NANOPARTICLE APPLICATIONS

Nanoparticles are usually produced from bulk metallic and non-metallic materials. The metallic nanoparticles include copper, aluminum, silver, iron and gold, whereas non-metallic nanoparticles used are alumina, copper oxide, titania, iron oxide, zinc oxide, carbon nanotubes, glass, polymers, etc. These nanoparticles exhibit extraordinary physical, mechanical, thermal, electrical, catalytic, anti-bacterial/antimicrobial, optical properties, electron transport, semiconducting property, ferromagnetism, giant magneto resistance, luminescence, ferroelectric property, etc. (Gleiter, 1992; Hosokawa et al., 2012; Sahu et al., 2017). Because of their extraordinary properties, the nanoparticles are used in interdisciplinary emerging applications. The brief view of the diverse applications of nanoparticles is shown in Table 3.3 (Wen et al., 2011; Kathiravan et al., 2010; Garg et al., 2008; Drelich et al., 2011; Kalpowitz et al., 2010; Sindhu et al., 2007; Timofeeva et al., 2007; Hemalatha et al., 2011; Lee and Kim, 2011; Taylor et al., 2013; Saidur et al., 2011; Kassaee and Buazar, 2009; Rai et al., 2006; Kulkarni et al., 2008; Basha and Anand, 2011; Pfeifer et al., 2005; Johnston and Wilcoxon, 2012; Park et al., 2000; Sahu et al., 2019; Guo et al., 2002; Lynch and Sondergaard, 2009; Haidar, 2009).

TABLE 3.3 Diverse applications of nanoparticles

• Automobile cooling and lubrication systems	• Machine tool coolants and lubricants
• MEMS devices (cooling of tiny passages, micro-scale patterning through inks, etc.)	• Space rocket propellants
• Defense explosives formulation	• Lithium ion batteries
• Fuel cells (as catalyst and active anode material)	• Heat exchangers and evaporators
• Biomedical (cancer therapy, drug delivery, protection of food from bacteria, etc.)	• Nuclear reactors (wash coating, channels cooling, etc.)
• Optical glasses (as thin film coatings)	• Textiles (face masks, clothes, etc.)
• Solar cells	• Solar collection systems (dish/power towers and flat plate collectors)
• Automobile cooling and lubrication systems	• Automobile cooling and lubrication systems

• Diesel generator jacket water coolant system	• Electrical motor windings
• Electrical generator windings	• Domestic refrigerators and chillers
• Cameras and displays	• Sodium lamps (as envelope materials)
• Heat engines (harmful pollutants reduction and knock detection)	• Waste heat collectors
• Electrical contact systems such as contactors, compliant electrical connectors, slip rings and rolling contacts	• Cosmetics
• Boiler flue gas temperature reduction	• Transformers
• Marines (to reduce bio-fouling in ship hulls, etc.)	• New sensors (for improving petroleum exploration)
• Water purification	• Photothermal hyperthermia
• Paints	• Conductive optoelectronics, etc.

3.4 NANOPARTICLES PRODUCTION TECHNIQUES

Nanoparticles are usually produced through the physical and chemical routes. In the physical route, nanoparticles are produced by either mechanical crushing of the bulk material or evaporation of the bulk material followed by its condensation and nucleation. But, in the chemical route, nanoparticles are produced either by reducing metallic salt solutions in the presence of capping agents or by in-flight reaction of a liquid containing material precursors or by evaporating/condensing metal oxides and non-metallic materials in the presence of oxygen/nitrogen/light hydrocarbon. The broad classification of nanoparticles production using physical and chemical routes, but not limited to, is shown in Figure 3.3 and their mechanisms are explained briefly below.

3.4.1 Ball Milling

Ball milling is a physical route approach to generate nanoparticles by applying mechanical energy on the bulk solid materials, which break the bonding between

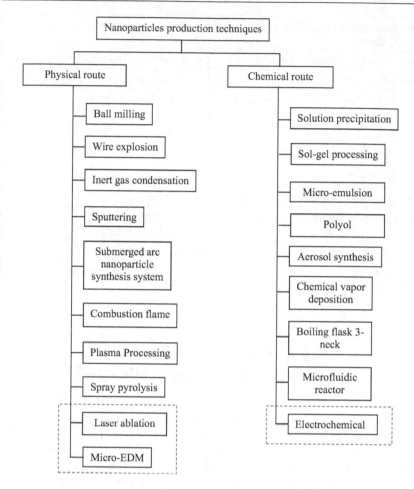

FIGURE 3.3 Classification of the mechanism of nanoparticle generation.

atoms and molecules. In this process, small balls are allowed to rotate inside a drum and drop due to gravitational force onto a solid enclosed in the drum, which breaks down the structure into crystallite nanoparticles. This is an established technology, which is simple and cheap (Wilson et al., 2002), but its main disadvantage is the contamination of the particles resulting from the grinding media.

Other disadvantages of this process include lack of control on the particle size distribution, highly polydispersed size distribution, introduction of internal stresses and partially amorphous state of the powder. However, this is the preferred technique for generating metal oxides and intermetallic compounds.

For example, $LiFePO_4$ nanoparticles of a mean diameter of 50 nm were synthesized by this process (Xinping et al., 2006).

3.4.2 Wire Explosion

In this technique, a high-voltage high-current pulse is supplied to the thin metal wires surrounded by an air environment or inert gas atmosphere. This results in vaporization and ionization of the wire. The plasma thus generated from the vaporized and ionized wire material expands outward due to high internal pressure and is cooled by collision with air or inert gas surrounding. Finally, the condensed vapor forms the nanosized particles. The advantages of this process are control over the size of the nanoparticles and very low energy consumption (Sarathi et al., 2007; Ivanov et al., 1995).

3.4.3 Inert Gas Condensation

The inert gas condensation technique involves evaporation of metal (by resistive heating, radio frequency heating and electron beam heating) in a chamber that is evacuated to a very high vacuum and then backfilled with low-pressure inert gas. The evaporated metal atoms collide with the gas atoms, lose their kinetic energy and condense to small discrete particles. Considerable control on the particle shape and size can be exercised in this technique (Gleiter, 1989; Swihart, 2003).

3.4.4 Sputtering

Sputtering is a physical vapor deposition process by which atoms in a solid target material are expelled into the gas phase due to the impact of an atom or ion on the material as a result of the momentum transfer. When ions of a suitable substance accelerated to high energies are directed toward a surface, atoms and clusters, both neutral and ionic, are expelled. The ratio of atoms to clusters and ions to neutrals produced depends on the mass and energy of the projectile ion and a variety of other experimental parameters. Unlike other vapor-phase techniques, there is no melting of the material takes place in this technique. The disadvantage of this technique is that it usually produces a small amount of clusters whose frequency distribution decreases exponentially with increasing cluster size. In a few cases, notably carbon and silicon, sputtering yields broader distributions of cluster ions (Gleiter, 1989; Suryanarayana and Koch, 2000).

3.4.5 Submerged Arc Nanoparticle Synthesis System (SANSS)

In submerged arc nanoparticle synthesis system, nanoparticles are generated in a dielectric medium. The metal is heated by an electric arc, which vaporizes the metal that is cooled by the surrounding dielectric fluid. The vapor condenses and forms nanoparticles. The advantage of this process is easy to control over the process parameter and the drawback being low yield (Chang et al., 2005).

3.4.6 Combustion Flame

In this technique, a steady flat flame is generated by burning a fuel-oxygen mixture. The chemical precursors, introduced along with combustibles, experience rapid thermal decomposition in the hot zone of the flame. Since the temperature profile, gas-phase residence time and precursor concentration are consistent across the entire surface of the burner, the effect is to generate monodispersed particles. The continuous stream of nanoparticle clusters generated from the combustion zone is quenched and the powder is collected (Swihart, 2003; Karasev et al., 2004).

3.4.7 Plasma Processing

In this technique, the precursor material is micron-sized particles. To generate very fine particles, the coarse particles were injected into a reaction chamber operating at atmospheric pressure and were subjected to extended arc plasma using graphite electrodes. The typical parameters used during plasma processing are an arcing current of 300 A, a load voltage of 50 V and argon plasma-forming gas. The particles are collected from the cold wall of the reaction chamber (Wilson et al., 2002).

3.4.8 Spray Pyrolysis

Spray pyrolysis is basically a solution process, which has been widely used in the preparation of metal and metal oxide powder. In this technique, a starting solution is typically prepared by dissolving the metal salt of the product in the solvent. From the starting solution, the liquid droplets are atomized and these micro-sized liquid droplets of precursor or precursor mixture are introduced into the furnace. Finally, they are converted into solid particles through heating.

In practice, spray pyrolysis involves several steps – (i) generating micro-sized droplets of liquid precursor or precursor solution, (ii) evaporation of solvent, (iii) condensation of solute, (iv) decomposition and reaction of solute and (v) sintering of solid particles (Messing et al., 1993; Gurav et al., 1993). However, the main disadvantage of this technique is that a porous particle is produced in the case of a high heating rate or a large droplet size. These morphological conditions are undesirable for most applications.

3.4.9 Solution Precipitation

The precipitation of solids from a metal ion containing solution is one of the most frequently employed generation processes for nanoparticles. In the precipitation technique, an inorganic metal salt is dissolved in water, then metal ions from metal hydrate species in water and these are hydrolyzed by adding a base solution. The hydrolyzed species condense with each other forming metal hydroxide precipitate. This precipitate is filtered out and dried, which is subsequently calcined to obtain the final crystalline nanoparticles. The advantage of this process is that it is economical, but the main drawback is the inability to control the size of the particles and their subsequent aggregation (Gleiter, 1989).

3.4.10 Sol-Gel Processing

The sol-gel process is one of the most popular solution processing techniques used primarily for the generation of metal oxide nanoparticles. In this process, a reactive metal precursor is hydrolyzed with water and the hydrolyzed contents are allowed to condense with each other to form precipitates of metal oxide nanoparticles. The precipitate is subsequently washed and dried, and then calcined at an elevated temperature to form crystalline metal oxide nanoparticles (Gleiter, 1989; Wilson et al., 2002). Some of the disadvantages associated with this process are costly metallic precursors and sensitivity for atmospheric conditions.

3.4.11 Micro-Emulsion

Micro-emulsion is a dispersion of fine liquid droplets of an organic solution in an aqueous solution. Such a micro-emulsion system can be used for the synthesis of nanoparticles. The chemical reactions can take place either at the interfaces between the organic droplets and aqueous solution, when reactants are introduced separately into two non-mixable solutions or inside the organic droplets when all the reactants are dissolved into the organic droplets (Cao, 2004).

3.4.12 Polyol Technique

This is a technique for preparing monodispersed, non-agglomerated nanoparticles, in which heating a suitable inorganic/organic metallic salt in polyol (i.e., alcohol having multiple hydroxyl groups) gives rise to metal particles, and polyol acts as a solvent and reducing agent (Figlarz et al., 1985). For example, polyols can effectively reduce metal ions to prepare sub-micrometer size particles of gold, silver and copper under microwave irradiation, and the approach is referred to as the microwave polyol process (Fievet et al., 1989). The drawback of this process is that the solution of metallic salt should be heated to its boiling point and kept under refluxing conditions for a long time (Zhu et al., 2004).

3.4.13 Aerosol Synthesis

In this technique, a liquid precursor is first prepared. The precursor can be a simple mixture solution of desired constituent elements or a colloidal dispersion. Such a liquid precursor is then mystified to make a liquid aerosol, i.e., a dispersion of uniform droplets of liquid in a gas, which may simply solidify through evaporation of solvent or further react with the chemicals that are present in the gas. The resulting particles are spherical and their size is determined by the size of the initial liquid droplets and the concentration of the solid. Aerosols can be relatively easily produced by sonication or spinning (Cao, 2004; Kalpowitz et al., 2010).

3.4.14 Chemical Vapor Condensation

Chemical vapor condensation is a vaporization technique where the starting material and the final nanoparticles have different compositions because of the chemical reaction occurring between the vapor and other components of the system during vaporization and condensation. The metal vapor is generated by heating metal carbonyl, organics, hydrides or chlorides in a chamber and the vapor is rapidly condensed in another cold chamber by carrying it with the help of a carrier gas. The nanoparticles are extracted from the aerosol formed by means of filters or electrostatic precipitators (Gleiter, 1989; Swihart, 2003).

3.4.15 Boiling Flask-3-Neck

In a boiling flask-3-neck technique, the diluted reactants (e.g., tetra amino copper hydroxide $[Cu(NH_3)_4]\cdot(OH)_2$ solution and hydrazine hydrate solution) are mixed under a magnetic stirrer at room temperature. The mixed solution undergoes a chemical reaction yielding a fast reduction rate of metallic ions to pure metallic nanoparticles. The size of the particles produced by this technique is usually difficult to control and the size distribution is relatively broad (Zhang et al., 2010).

3.4.16 Microfluidic Reactor

In a microfluidic synthesis technique, the reactants (e.g., tetra amino copper hydroxide solution, hydrazine hydrate and trisodium phosphate solution) were preloaded in 1 ml gastight syringes to which microfluidic reactors were connected. The preloaded reactants are introduced into the microfluidic reactors by a syringe pump. The mixed reactant that flows through the reaction channel undergoes a chemical change and results in the formation of the colloidal suspension of nanoparticles. The advantage of this technique is that it eliminates the local variations in the reaction conditions such as concentration and temperature (Zhang et al., 2010).

3.5 EXPLOITATION OF NANOTECHNOLOGY CONCEPT IN ADVANCED MACHINING

Advanced machining techniques are those techniques that are used to produce miniature features by removing materials from the workpiece using mechanical/thermal/chemical and electrochemical source of energy with no constraint on the size of the workpiece. In these techniques, the material is removed in the form of debris at the micro or nano level. The advanced machining techniques based on different sources of energy and their principle of material removal were outlined in Section 1.6, Chapter 1.

The important features of these techniques include very low input energy can be used in terms of micro Joule resulting in low unit removal of material, involves micron-size tool and a low range of operating parameters, able to machine general purpose and exotic materials that are difficult to machine by conventional machining and produce intricate micro features and able to meet the surface quality requirement at the nano level. A detailed explanation of these techniques is available in various literature (Jain 2010; Ghosh and Mallik, 2010; Payal and Sethi, 2003; Lee et al., 2016; Masuzawa, 2000) and therefore not covered in this book.

It has been known that advanced machining techniques have found widespread applications in many industrial domains. The applications include cooling holes in aircraft turbine blades, filters for food processing and textile industries, micro holes in fuel injection nozzles, spinneret holes for synthetic fibers, catheters, needles and other medical devices, fiber-optics, micro-optoelectronic components, micro-mechatronic actuator parts, microchannels in nuclear reactors, MEMS devices and microfluidic devices. These miniature features are obtained by removing a tiny bit of materials in the form of debris leaving behind a crater on the surface of workpiece. The debris formed is treated as unwanted materials and is disposed off to the environment. The debris formed is of quite small enough and becomes a challenge for the researchers to concentrate on the debris and to collect, test, reveal the structure of debris and its possible application as nanoparticles in the field of nanotechnology. Recently, some of the advanced machining techniques like laser ablation, electrochemical micromachining (ECMM) and micro-electrical discharge machining (micro-EDM) are explored for the production of nanoparticles and are briefly discussed in the following sections.

3.5.1 Pulsed Laser Ablation

In pulsed laser ablation (also called laser beam micromachining) technique, a pulsed laser beam of high intensity power is concentrated on a target material. As a result, ablation of the material (i.e., melting and evaporation) occurs and the vapor cloud of materials is tightly confined both spatially and temporally. Thus, nanoparticles are produced. This technique is simple and flexible, and the synthesis of nanoparticles can take place in any arbitrary liquids. However, the disadvantages of this process are very low yield of nanoparticles and the mechanism of laser-matter interactions is complex. This technique is used for synthesizing metal and metal oxide nanoparticles (Swihart, 2003; Tilaki et al., 2007; Sahu et al., 2019).

3.5.2 Electrochemical (or Electrochemical Micromachining) Technique

In electrochemical technique or ECMM technique, an electric current is passed between two electrodes – cathode and anode separated by an electrolyte and nanoparticles/nanostructures are produced by anodic dissolution of the workpiece. The process of producing nanoparticles/nanostructures using ECMM technique is detailed below.

Figure 3.4 shows the schematic diagram of the ECMM cell. The cell usually consists of a tool (cathode) and workpiece (anode) connected to a DC power supply and is separated by a small inter-electrode gap. The workpiece is supported by a workpiece holder built inside a work tank. A tool feed control unit was used to control the tool feed through an actuator to maintain a constant inter-electrode gap during machining. Tool feed control can be achieved based on the average gap signal (may be back pressure/voltage signal) as the feedback signal. The required voltage is supplied between the electrodes using a DC power supply. The tool moved toward the workpiece until it reaches a set inter-electrode gap. The prepared electrolyte solution of suitable concentration contained in an electrolyte tank is pumped at high pressure into the tool (nozzle shaped). While the flow of the electrolyte is

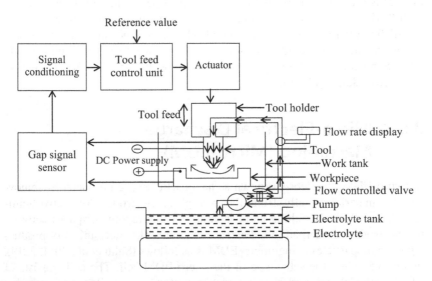

FIGURE 3.4 Schematic of the ECMM cell.

precisely controlled by a valve operated through a micro-controller. The high-velocity electrolyte from the tool impinged on a certain region of work-piece and the electric current passed between the two electrodes used to remove the material from the workpiece in the form of debris by ionic displacement based on an electrolysis phenomenon leaving behind an ultrafine crater on the workpiece surface. The debris thus generated was dispersed in the solution and then collected and centrifuged. Finally, the solid debris as nanoparticles was produced. At the same time when the workpiece is eroded, the inter-electrode gap changes and this gap signal is sensed, signal conditioned and compared with the reference value in the tool feed control unit corresponding to the set inter-electrode gap. The error signal is corrected by a correction element in the tool feed control unit supplied to the actuator, which in turn would feed the tool and controls the inter-electrode gap, so that the material removal would carry on efficiently at each location without any disturbance. Moreover, researchers are focused on the ultrafine craters formed on the workpiece surface due to anodic dissolution as nanostructures which can be used for various industrial applications. These nanostructures can also ascertain the nanoscale accuracy of ECMM. The use of suitable stabilizers plays an important role during the process in order to prevent agglomeration/aggregation of nanoparticles produced and also to aid in the rapid removal of particles from the cathode surface to the electrolyte solution. This technique involves low cost, simple operation, high yield rate and high flexibility, but its disadvantages include chemicals often used are toxic and difficult to remove from the colloid. This technique can be used for producing metal, metal oxide and semiconductor nanoparticles (Taylor et al., 2013; Sahu et al., 2019; Dongping et al., 2016; Roldan et al., 2013; Dongping et al., 2017; Jain et al., 2019).

3.5.3 Micro-Electrical Discharge Machining (Micro-EDM)

In micro-EDM, the debris (known as nanoparticles) are produced by removing the material usually from the workpiece electrode (conductive/semi-conductive) through melting and evaporation using repetitive sparks between the electrodes submerged in a dielectric medium. The mechanism of producing the nanoparticles using micro-EDM is as follows (Sahu et al., 2014, 2019). Figure 3.5 shows the schematic of the micro-EDM cell. The cell consists of a tool (cathode) and workpiece (anode) separated by a small spark gap and is submerged in a dielectric medium contained in an ultrasonicator. The ultra-sonicator produces ultrasonic vibration to the dielectric medium in order to

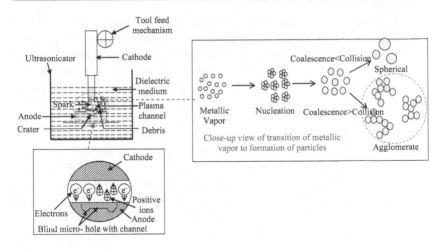

FIGURE 3.5 Schematic of the micro-EDM cell.

overcome agglomeration of debris (particles) produced during machining. The ultrasonicator can also assist in the removal of debris from the spark gap zone proficiently and avoid the accumulation of debris in the spark gap so that stable spark discharges occur and efficient machining can be accomplished. A tool feed mechanism can be used for tool feed control in order to maintain a constant spark gap between the electrodes.

The required open circuit voltage is applied between the electrodes using a regulated DC power supply. The tool electrode moved toward the workpiece electrode until the gap between the electrodes should reach a value equivalent to the set spark gap. When the voltage builds up to some predetermined value, electrons break loose from the tool (cathode) and are accelerated toward the workpiece (anode). During their travel within the spark gap, the electrons collide with the neutral molecules of dielectric fluid causing ionization and result in the formation of a plasma channel between the two electrodes. Since the electrical resistance of the plasma channel is very less, all of a sudden a large number of electrons will move from the tool to the workpiece and positive ions from the workpiece to the tool. As a result, a spark discharge takes place. Due to the spark discharge, a localized intense heat is produced which increases the workpiece temperature in a narrow zone, resulting in melting and vaporization of the work material. Thus, the vaporized material condensed followed by nucleation and growth and resulted in the production of debris as nanoparticles leaving behind a small crater on the workpiece surface.

The mechanism of production of debris (nanoparticles) in the dielectric medium is shown in Figure 3.4 (a close-up view of the transition of metallic vapor to the formation of particles). The metallic vapor molecules that could be produced due to melting and evaporation during micromachining will collide with the atoms of the dielectric medium at room temperature and become cooled. As a result, metallic vapor particles temperature could be suddenly decreased and reaches a super-cooled state. At this state, the vapor particle will start to condense and nucleate. The transformation of phase will start with the formation of homogeneous nucleation of particles and then growth will take place till the complete phase change occurs. This indicates that after nucleation, particles generally continue to grow by gaining more atoms from the metallic vapor. Moreover, growth will occur when the particles collide with each other and then coalesce to form larger units. At high concentrations, the particles growth could be governed by the coalescence and collision rates. If the rate of coalescence is more than the collision rate, then collided particles would be clustered due to van der Waals' forces of attraction and thus agglomerate structures could be produced. If the collision rate is faster, then primary spherical particles could be produced. To avoid agglomeration, capping agents were used in the dielectric medium. The capping agent molecules dissolved in the dielectric medium and during the production of debris (particles) the capping molecules could arrange themselves at the interface between solid particles and liquid medium because of their amphiphilic nature. This leads to the formation of a barrier that could prevent close contact between particles and inhibits agglomeration.

As the workpiece eroded, the spark gap increases and the gap voltage also changes. The gap voltage signal is supplied to the tool feeding unit through the tool feed control unit which in turn feed the tool and controls the gap. Accordingly, the machining at each point could continue without any interruption. This technique is simple, versatile and cost-effective and can produce nanoparticles of approximately uniform size in the form of colloid/without colloid with high yield in a single step by closely controlling the process parameters, but the diffusion-limited growth control is still a challenging issue. This technique is used for producing metal, metal oxide and semiconductor nanoparticles.

3.5.4 Micro-Electro Chemical Discharge Machining (Micro-ECDM)

Micro-electro chemical discharge machining (micro-ECDM) also known as electrochemical spark machining (ECSM) is a hybrid advanced machining technique currently explored for the production of conducting and non-conducting nanoparticles. Micro-ECDM technique can produce nanoparticles via the physicochemical route (i.e., through the integration of micro-EDM and ECMM principle). This means that in this technique, nanoparticles are produced by removing the material usually from the workpiece in the form of debris through the combined principle of melting and evaporation, and ionic displacement (chemical etching/anodic dissolution) of workpiece using electrical discharges between the electrodes immersed in an electrolyte solution (Singh et al., 2016; Bishwakarma and Das, 2020; Bindu and Somashekhar, 2021).

In micro-ECDM technique, for production of conducting (e.g., metal/metal oxide) nanoparticles, the electrodes – tool (cathode) and workpiece (anode) – were connected to the pulse DC power supply. They were immersed in an electrolyte solution (alkali solution like KOH, NaOH solution, etc.) contained in an electrolyte tank (usually made of acrylic sheet) and separated by a small gap. The electrolyte tank supported on the magnetic stirrer could provide agitation to the electrolyte to avoid the agglomeration of debris produced during machining (Figure 3.6a). An additional power supply unit was used to control the actuator and electrolyte flow control units. The constant gap between the electrodes could achieve by an actuator controlled by the software. The experimental parameters like voltage, current, frequency and duty factor were set. When pulsed DC voltage was applied, initially H_2 gas bubbles of smaller radius were formed due to the initiation of electrochemical reaction. With an increase in voltage below the critical voltage level, there resulted in the formation of more H_2 bubbles. These bubbles joined with other formed bubbles that could lead to the formation of a larger radius of H_2 gas bubbles. At the same time due to Ohmic heating of the electrolyte attributed to increased voltage, the electrolyte evaporates in the vicinity of the tool that resulted in the formation of vapor bubbles. These vapor bubbles combined with the H_2 gas bubbles of larger size, and thus a gaseous film/layer was formed around the surface of the tool electrode. Because of the gaseous film formation, there was an obstruction between the cathode and the electrolyte. This could result in an increase in current density within the gas film. Further increase in voltage above the critical voltage level and high current density could lead to ionization of gas film and subsequent movement of electrons and holes between the electrodes. This could enable the spark discharges. Due to the spark, the workpiece material is heated to melt and evaporate, and simultaneously anodic dissolution (chemical etching) of the workpiece also takes place. As a result, the material removed from the workpiece condensed to form debris (i.e., nanoparticles) via nucleation and growth. The electrolyte colloidal solution of nanoparticles was centrifuged and the precipitate formed

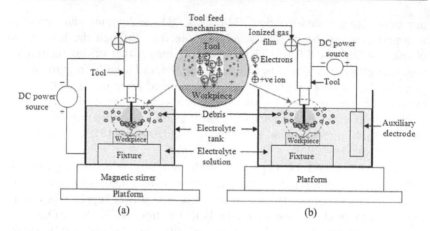

FIGURE 3.6 Schematic of the micro-ECDM cell: (a) for production of conducting nanoparticles and (b) for production of non-conducting nanoparticles.

was washed repeatedly with deionized water and some organic liquids again by centrifuging until the electrolyte separated from the precipitate. Finally, the precipitate was dried in an oven to form solid nanoparticles (Singh et al., 2016, 2018; Bishwakarma and Das, 2020).

Further, in micro-ECDM technique, for production of non-conducting (e.g., glass or polymers) nanoparticles, the workpiece material (electrically non-conductive) was placed near the tool (cathode) and an auxiliary electrode (anode) was placed at a certain small distance (about 25–50 mm) from the tool (Figure 3.6b). The workpiece and electrodes were immersed in an electrolyte solution (e.g., standard alkali solutions) contained in a beaker. The electrodes were connected to the pulse DC power supply. The tool moved toward the workpiece using a tool feed control unit. The experimental parameters like voltage, current, frequency, duty factor and feed rate were set. The mechanism of formation of sparks was similar to that as explained above for the production of conducting nanoparticles. But, in this case, the workpiece was placed near the tool (chemically non-reactive) where sparking occurs. The repetitive sparks heated to melt and evaporate the workpiece and simultaneously chemical etching of the workpiece also occur. As a result, the material removed from the workpiece condensed to form debris (i.e., nanoparticles) via nucleation and growth. The electrolyte colloidal solution of nanoparticles was washed with deionized water several times by centrifuging until the electrolyte separated from the precipitate. Finally, the precipitate was dried in an oven to form solid non-conducting nanoparticles. As the material was removed, the tool fed toward the workpiece using the tool feeding unit and maintained a constant gap (Bindu and Somashekhar, 2021).

3.6 CHALLENGES OF NANO-BASED MANUFACTURING

The challenges faced by the researchers in nano-based manufacturing techniques, especially in advanced machining techniques, include the following:

- Production of uniform size bare nanoparticles with controlled morphology and at high yield rate,
- Diffusion-limited growth control to accomplish uniform size distribution,
- Long-term stability of colloidal nanoparticles,
- Reduction in the anomaly in thermal conductivity of colloids and
- Production of hybrid nanoparticles with approximate uniformity in sizes.

These challenges will provide a wealth of knowledge in making the advanced machining techniques an ideal nano-based manufacturing technique for various promising applications.

3.7 SUMMARY

In this chapter, the introduction to nanotechnology followed by the science and application of nanoparticles were briefly presented. The variety of techniques used for the production of nanoparticles were discussed. Further, the recent exploration of advanced machining techniques for the production of nanoparticles was detailed and the challenges of nanoparticle-based manufacturing were briefly outlined.

REFERENCES

Al-Kayiem, H., Lin, S., and Afolabi, L. (2013) Review on nanomaterials for thermal energy storage technologies. *Nanoscience & Nanotechnology-Asia*, 3(1), 60–71.
Basha, J.S. and Anand, R.B. (2011) An experimental study in a CI engine using nano-additive blended water-diesel emulsion fuel. *International Journal of Green Energy*, 8 (3), 332–348.

Bindu, M.J. and Somashekhar, S.H. (2021) Generation and characterization of borosilicate glass nanoparticles using in-house developed μ-ECDM Setup, Silicon, https://doi.org/10.1007/s12633-021-00986-9.

Bishwakarma, H. and Das, A.K. (2020) Synthesis of zinc oxide nanoparticles through hybrid machining process and their application in supercapacitors, *Journal of Electronic Materials*, 49, 1541–1549. https://doi.org/10.1007/s11664-019-07835-x.

Chang, H., Tsung, T.T., and Lo, C.H. (2005) A study of nanoparticle manufacturing process using vacuum submerged arc machining with aid of enhanced ultrasonic vibration. *Journal of Materials Science*, 40, 1005–1010.

Cao, G. (2004) *Nanostructures and Nanomaterials---Synthesis, Properties and Applications*, UK, Imperial College Press.

Darweesh, H.H.M., (2018) Nanomaterials---classification and properties-Part I. *Nanoscience*, 1, 1, 1–11.

Dongping, Z., Lianhuan, H., Jie, Z., Kang, S., Jian-Zhang, Z., Zhao-Wu, T., and Zhong-Qun, T. (2016) Confined chemical etching for electrochemical machining with nanoscale accuracy, *Accounts of Chemical Research*, 49(11), 2596–2604.

Dongping, Z., Lianhuan, H., Jie, Z., Quanfeng, H., Zhao-Wu, T., and Zhong-Qun, T. (2017) Electrochemical micro/nano-machining---principles and practices, Electrochemical micro/nano-machining---principles and practices, *Chemical Society Reviews*, 46(5), 1526–1544.

Drelich, J., Li, B., Bowen, P., Hwang, J., Mills, O., and Hoffman, D. (2011) Vermiculite decorated with copper nanoparticles---antibacterial hybrid material. *Applied Surface Science*, 257, 9435–9443.

Fievet, F., Lagier, J.P., and Figlarz, M. (1989) Preparing monodisperse metal powders in micrometer and submicrometer sizes by the polyol process. *Materials Research Bulletin*, 14, 29–34.

Figlarz, M., Fievet, F., and Lagier, J.P. (1985) Process for the Reduction of Metallic Compounds by Polyols, and Metallic Powders Obtained By This Process, U.S. Patent 4539041.

Garg, J., Poudel, B., Chiesa, M., Gordon, J.B., Ma, J.J., Wang, J.B., Ren, Z.F., Kang, Y.T., Ohtani, H., Nanda, J., McKinley, G.H., and Chen, G. (2008) Enhanced thermal conductivity and viscosity of copper nanoparticles in ethylene glycol nanofluid. *Journal of Applied Physics*, 103, 074301/1–074301/6.

Ghosh, A. and Mallick, A.K. (2010) *Manufacturing Science*, 2nd edition, East-West Press, Bangalore.

Gleiter, H. (1989) Nanocrystalline materials. *Progress in Materials Science*, 33, 223–315.

Gleiter, H. (1992) Materials with ultrafine microstructures---retrospectives and perspectives. *Nanostructured Materials*, 1, 1–20.

Gleiter, H. (2000) Nanostructured materials---basic concepts and microstructure. *Acta Materialia*, 48 (1), 1–20.

Guo, D., Xie, G., and Luo, J. (2013) Mechanical properties of nanoparticles---basics and applications. *Journal of Physics D---Applied Physics*, 47(1), 013001.

Guo, K., Pan, Q., Wang, L., and Fang, S. (2002) Nano-scale copper-coated graphite as anode material for lithium-ion batteries. *Journal of Applied Electrochemistry*, 32, 679–685.

Gurav, A., Kodas, T., Pluym, T., and Xiong, Y. (1993) Aerosol processing of materials. *Aerosol Science and Technology*, 19 (4), 411–452.

Haidar, J. (2009) Synthesis of Al nanopowders in an anodic arc. *Plasma Chemistry and Plasma Processing*, 29, 307–319.

Heinz, H., Pramanik, C., Heinz, O., Ding, Y., Mishra, R. K., Marchon, D., Flatt, R. J., Irin, E. L., Llop, J., Moya, S., and Ziolo, R.F. (2017) Nanoparticle decoration with surfactants---Molecular interactions, assembly, and applications. *Surface Science Reports*, 72(1), 1–58.

Hemalatha, J., Prabhakaran, T., and Nalini, R.P. (2011) A comparative study on particle-fluid interactions in micro and nanofluids of aluminium oxide. *Microfluid Nanofluid*, 10, 263–270.

Ivanov, V., Kotov, Y.A., Samatov, O.H., Bohme, R., Karow, H.V., and Schumacher, G. (1995) Synthesis and dynamic compaction of ceramic nanopowders by techniques based on electric pulsed power. *Nanostructured Materials*, 6, 287–290.

Jain, N. K., Pathak, S., and Alam, M. (2019) Synthesis of copper nanoparticles by pulsed electrochemical dissolution process, *Industrial and Engineering Chemistry Research*, 58(2), 602–608.

Jain, V.K. (2010) *Introduction to Micromachining*, Narosa Publishing House Pvt. Ltd., India

Jalill, R.D.A., Nuaman, R.S., and Ahmed, N.A. (2016) Biological synthesis of Titanium Dioxide nanoparticles by Curcuma longa plant extract and study its biological properties, *World Scientific News*, 49(2), 204–222.

Johnston, R.L. and Wilcoxon, J.P. (2012) *Frontiers of Nanoscience*, Volume 3, Elsevier Science, The Netherlands.

Kalpowitz, D.A., Jouet, R.J., and Zachariah, M.R. (2010) Aerosol synthesis and reactive behavior of faceted aluminum nanocrystals. *Journal of Crystal Growth*, 312, 3625–3630.

Kassaee, M.Z and Buazar, F. (2009) Al nanoparticles---impact of media and current on the arc fabrication. *Journal of Manufacturing Processes*, 11, 31–37.

Karasev, V.V, Onischuk, A.A., Glotov, O.G., Baklanov, A.M., Maryasov, A.G., Zarko, V.E., Panfilov, V.N., Levykin, A.I., and Sabelfeld K.K. (2004) Formation of charged aggregates of Al_2O_3 nanoparticles by combustion of aluminum droplets in air. *Combustion and Flame*, 138, 40–54.

Kathiravan, R., Kumar, R., Gupta, A., and Chandra, R. (2010) Preparation and pool boiling characteristics of copper nanofluids over a flat plate heater. *International Journal of Heat and Mass Transfer*, 53, 1673–1681.

Kulkarni, D., Vajjha, R., Das, D., and Oliva, D. (2008) Application of aluminum oxide nanofluids in diesel electric generator as jacket water coolant. *Applied Thermal Engineering*, 28, 1774–1781.

Lee, C.M., Woo, W.S., Kim, D.H., Oh, W.J., and Oh, N.S. (2016) Laser-assisted hybrid processes---A review. *International Journal of Precision Engineering and Manufacturing*, 17, 257–267.

Lee, H.M. and Kim, Y.J. (2011) Preparation of size-controlled fine Al particles for application to rear electrode of Si solar cells. *Solar Energy Materials and Solar Cells*, 95, 3352–3358.

Luther, W. (2004) *Industrial Application of Nanomaterials*, Future Technology Division, Germany.

Lynch, W.A. and Sondergaard, N.A. (2009) Liquid additives to improve conductivity in electric contacts. *Proceedings of Fifty-Fifth IEEE Holm Conference on Electrical Contacts*, 78–86.

Maier, W.F. (1990) Transition Metal Oxides---Surface Chemistry and Catalysis. (Reihe---Studies in Surface Science and Catalysis, Vol. 45). Von H. H. Kung. Elsevier, Amsterdam 1989. X, 282 S., geb. HFl. 215.00. — ISBN 0–444–87394-5, *Angewandte Chemie, Wiley,* 102(8), 965–966.

Mansoori, A.G. (2005) *Principles of Nanotechnology---Molecular-Based Study of Condensed Matter in Small Systems,* World Scientific, Singapore.

Masuzawa, T. (2000) State of the art of micromachining. *Annals of the CIRP,* 49, 473–488.

Messing, G.L., Zhang, S.C., and Jayanthi, G.V. (1993) Ceramic powder synthesis by spray pyrolysis. *Journal of the American Chemical Society,* 76, 2707–2726.

Park, S., Gorte, R.J., and Vohs, J.M. (2000) Applications of heterogeneous catalysis in the direct oxidation of hydrocarbons in a solid-oxide fuel cell. *Applied Catalysis A---General,* 200, 55–61.

Payal, H.S. and Sethi, B.L. (2003) Non-conventional machining processes as viable alternatives for production with specific reference to electrical discharge machining. *Journal of Scientific and Industrial Research,* 62, 678–682.

Pfeifer, P., Schubert, K. and Emig, G. (2005) Preparation of copper catalyst washcoats for methanol steam reforming in microchannels based on nanoparticles. *Applied Catalysis A---General,* 286, 175–185.

Poole, C.P. and Owens, F.J. (2003) *Introduction to Nanotechnology,* Wiley Interscience, USA.

Rai, A., Park, K., Zhou, L., and Zachariah M.R. (2006) Understanding the mechanism of aluminium nanoparticle oxidation. *Combustion Theory and Modelling,* 10 (5), 843–859.

Reghunadhan, A., Kalarikkal, N., and Thomas, S. (2018) Chapter 7- Mechanical Property Analysis of Nanomaterials, In *Characterization of Nanomaterials,* Mohan Bhagyaraj, S., et al., Ed. Woodhead Publishing, Sawston, 91–212.

Roldan, M.V., Pellegri, N., and Sanctis, O. (2013) Electrochemical method for Ag-PEG nanoparticles synthesis, *Journal of Nanoparticles,* 2013, Article ID 524150.

Saidur, R., Leong, K.Y., and Mohammad, H.A. (2011) A review on applications and challenges of nanofluids. *Renewable and Sustainable Energy Reviews,* 15, 1646–1668.

Sahu, R.K. and Hiremath, S.S., (2017) Synthesis of aluminium nanoparticles in a water/polyethylene glycol mixed solvent using μ-EDM. *IOP Conference Series---Materials Science and Engineering,* 225, 012257.

Sahu, R.K.; Hiremath, Somashekhar S. (2017) Electrical Discharge Machining (EDM)---Nanoparticle Generation. In Encyclopedia of Plasma Technology, 1st Ed.; Shohet, J., Ed.; CRC Press, Taylor & Francis, New York, 1, 432–442.

Sahu, R.K., Somashekhar S H. (2019) *Corona Discharge Micromachining for the Synthesis of Nanoparticles---Characterization and Applications* 1st Edition, Print ISBN -9780367224738; eBook ISBN – 9781000065404, DOI. 10.1201/9780429275036, CRC Press, Taylor & Francis, Boca Raton, New York.

Sahu, R.K., Somashekhar S.H., Manivannan, P.V., and Singaperumal, M. (2014) Generation and characterization of copper nanoparticles using micro-electrical discharge machining, *Materials and Manufacturing Processes,* 29 (4), 477–486.

Sahu, R.K. and Somashekhar S.H. (2019) Role of Stabilizers on Agglomeration of Debris during Micro-Electrical Discharge Machining, *Machining Science and Technology,* 23 (3), 339–367.

Sant, S.B. (2012) Nanoparticles---From Theory to Applications. *Materials and Manufacturing Processes*, 27(12), 1462–1463.

Santos, C.S.C., Gabriel, B., Blanchy, M., Menes, O., García, D., Blanco, M., and Arconada, N. (2015) Victor Neto Industrial Applications of Nanoparticles – A Prospective Overview, *Materials Today---Proceedings*, 2(1), 456–465.

Sarathi, R., Sindhu, T.K. and Chakravarthy, S.R. (2007) Generation of nano aluminium powder through wire explosion process and its characterization. *Materials Characterization*, 58 (2), 148–155.

Sindhu, T.K., Sarathi, R. and Chakravarthy, S.R. (2007) Generation and characterization of nano aluminium powder obtained through wire explosion process. *Bulletin of Materials Science*, 30 (2), 187–195.

Singh, P.K, Kumar, P., Hussain, M., Das, A.K. and Nayak, G.C. (2016) Synthesis and characterization of CuO nanoparticles using strong base electrolyte through electrochemical discharge process, *Bulletin of Materials Science*, 39(2), 469–478.

Singh, P.K, Shubham, Singh, N.K, Bishwakarma, H., Hussain, M., Das, A.K., and Prasad, B.H. (2018) Effect of annealing on silver oxide nano particle generated by electrochemical discharge machining, *Materials Today---Proceedings*, 5, 26804–26809.

Strambeanu, N., Laurentiu, D., Dan, D., and Mihai, L. (2015) Nanoparticles---Definition, Classification and General Physical Properties. In *Nanoparticles' Promises and Risks*, Lungu, M., Neculae, A., Bunoiu, M., and Biris, C., Eds., Springer, Cham, 3–8.

Suryanarayana, C. and Koch, C.C. (2000) Nanocrystalline materials-current research and future directions. *Hyperfine Interactions*, 130, 5–44.

Swihart, M.T. (2003) Vapor phase synthesis of nanoparticles. *Current Opinion in Colloid and Interface Science*, 8, 127–133.

Taylor, R., Coulombe, S., Otanicar, T., Phelan, P., Gunawan, A., Lv, W., Rosengarten, G., Prasher, R., and Tyagi, H. (2013) Small particles, big impacts---a review of the diverse applications of nanofluids. *Journal of Applied Physics*, 113, 011301/1–011301/19.

Tilaki, R.M., Irajizad, A., and Mahdavi, S.M. (2007) Size, composition and optical properties of copper nanoparticles prepared by laser ablation in liquids. *Applied Physics A---Materials Science and Processing*, 88, 415–419.

Timofeeva, E.V., Gavrilov, A.N., McCloskey, J.M., Tolmachev, Y.V., Sprunt, S., Lopatina, L.M., and Selinger, J.V. (2007) Thermal conductivity and particle agglomeration in alumina nanofluids---experiment and theory. *Physical Review E*, 76, 061203/1–061203/16.

Wen, J., Li, J., Liu, S., and Chen, Q. (2011) Preparation of copper nanoparticles in a water/oleic acid mixed solvent via two-step reduction method. *Colloids and Surfaces A---Physicochemical and Engineering Aspects*, 373, 29–35.

Wilson, M., Kannangara, K., Smith, G., Simmons, M., and Raguse, B. (2002) *Nanotechnology---Basic Science and Emerging Technologies*, Chapman and Hall, Australia.

Xinping, A., Hai, L., Xiaoyan, L., Qinlin, L., Bingdong, L., and Hanxi, Y. (2006) One-step ball milling synthesis of LiFePO4 nanoparticles as the cathode material of Li-ion batteries. *Wuhan University Journal of Natural Science*, 11(3), 687–690.

Zhu, H., Lin, Y., and Yin, Y. (2004) A novel one-step chemical method for preparation of copper nanofluids. *Journal of Colloid and Interface Science*, 277, 100–103.

Zhang, Y., Jiang, W., and Wang, L. (2010) Microfluidic synthesis of copper nanofluids. *Microfluid Nanofluid*, 9, 727–735.

Ultrafine Electronic Devices Manufacturing Techniques

4

4.1 SEMICONDUCTOR

Semiconductors are one of the electronic materials that behave in between conductors and insulators. They have higher electrical conductivity than insulators but lower conductivity than conductors at room temperature. The energy gap (band gap) of semiconductors lies in between 0 and 4 eV due to which they have moderate conductivity and carrier density at room temperature. They strongly absorb light, i.e., near IR and visible regions, and having energies above the energy gap. The semiconductor materials are Si, Ge, Ga, As, In, Sb, Se, etc., in which Si and Ge lie in the IV column of the periodic table known as elemental semiconductors. Further, Zn, Cd, Ga, In, As, Sb, Se and Te lie in II, III, V and VI columns of the periodic table, respectively. The combination of II and VI, III and V column atoms gives the binary compound, i.e., known as compound semiconductors. There are also ternary (GaAsP) and quaternary (InGaAsP) compounds that are exited in semiconductors (Swaminathan, 2017;

DOI: 10.1201/9781003203162-4

Groover, 2018). It is important to note that insulators at higher temperatures behave like semiconductors.

4.1.1 Semiconductor Doping

Doping in semiconductors is very important to make it electrically conductive. This can be achieved by the addition of elements, known as impurities. This process is known as doping. For example, if arsenic is added to silicon as an impurity, then it behaves like a conductor. This is decided through the valence of electrons. It is noted that one valence electron remains in arsenic after bonding with silicon. This left electron results in the flow of current. Further, if more arsenic atoms are added, then a large number of free electrons are generated, resulting in low resistance due to which free flow of current. Interestingly, if a few boron atoms are added with arsenic, then a few electrons are absorbed by boron, causing high resistance due to which low flow of current. Therefore, it can be said that controlling dopants can control any resistance. Using diffusion and ion implantation techniques can carry the doping in the semiconductor. By using these techniques, n-type or p-type semiconductors are manufactured (Swaminathan, 2017).

4.1.1.1 Diffusion Technique

In this process, the dopant atoms are placed on the surface of the semiconductor followed by heating at 800°C–1250°C. This heating is derived from the diffusion of dopant atoms into the semiconductor. The theory of diffusion is studied by Fick and gives a law known as Fick's law. The diffusion is categorized into intrinsic diffusion and extrinsic diffusion. It is a less costly and simple process, widely used to perform doping on wafers (Swaminathan, 2017; Groover, 2018).

4.1.1.2 Ion Implantation

In this process, dopant atoms are implanted into a semiconductor by using higher energy of the ion beam. This higher energy is transferred to dopant due to which it is penetrated up to a several micron depth on the semiconductor surface. The damage in the lattice is noticed due to the penetration of atoms. Heating at a moderate temperature can eliminate this damage. This process is known as annealing. This process is more expensive and complex. Further, it is an anisotropic process because the dopant is not speared much as compared with diffusion. This process assists the manufacture of self-aligned structures that significantly improve the performance of transistors.

4.2 IMPORTANCE OF SEMICONDUCTORS

Semiconductors are the most important materials in the fabrication of electronic devices. They became more important because their electrical properties can be changed by mixing a controlled amount of selected impurity atoms (known as dopants) with their crystal structures (Swaminathan, 2017). The dopants have either one more valence electron (n-type or negative dopant) or one less valence electron (p-type or positive dopant) than the atoms in the lattice. Presently, silicon is the most ideal material for the fabrication of semiconductor devices such as transistors and ICs. Compound semiconductors such as GaAs are widely used in the fabrication of laser diodes for compact disc player, LED (light emitting diode), satellite disc and cellular phone.

4.3 MATERIALS

Silicon is the most ideal material used in semiconductor industries. This is the group V element in the periodic table along with germanium. It is more stable in air even at elevated temperatures owing to the formation of a protective oxide film. It is a very brittle and hard crystalline solid with a dark-gray metallic luster and also the surface of silicon has shown more hydrophilic nature due to which adhesion is more on the surface. It has an octahedral crystal structure, the same as the diamond crystal structure formed by the merging of two FCC unit cells. It is tetravalent in nature and characterized by covalent bonding. Further, it is non-toxic, high melting point (1417°C) and insoluble in all ordinary solids. It has two forms of allotropes, i.e., brown amorphous and dark-gray crystalline. It has more advantages over germanium such as a large band gap (1.1 eV), operating up to 150°C and more suitable for electronic devices due to the easy formation of oxide (Madou, 2002; Bhushan, 2017; Groover, 2018; Petersen, 1982).

4.3.1 Silicon Wafer Fabrication

The fabrication of silicon wafer is carried out by the growth of a single crystal that is possible by using two techniques known as Czochralski (CZ) Crystal Growing and Floating Zone (May and Spanos, 2006; Groover, 2018).

The following sequence is used for the fabrication of silicon wafer. It is a single crystal growth → shaping → wafer slicing → wafer lapping → etching → polishing and cleaning → inspection and packaging.

4.3.1.1 CZ Crystal Growing

The crystal pulling method is also known as the Czochralski crystal growth process. In this process, a seed crystal of required diameter is dipped into the molten material of polysilicon and then slowly pulled at a rate of $10-20\,\mu m/s$ with a small rotation of 1 rev/s. Pulled material is started to solidify on seed crystals and the crystal structure of the seed is continued throughout. This process grows the single crystal of Si and Ge. The ingot of a single crystal is obtained typically 1 m length and 50–150 mm diameter.

4.3.1.2 Floating Zone

This method is widely used to develop the single crystal for fabrication of microelectronics devices. In this process, the polycrystalline silicon is kept on a single crystal and then moves slowly in the upward direction to pass through the induction coils, due to which the single crystal grows in the upward direction while maintaining its orientation. In this method, the regrowing silicon crystal rejects the dopant and stays in the liquid.

4.4 MEASUREMENT OF WAFER CHARACTERISTIC

After wafer fabrication, the quality of wafer is very important for the investigation of enhanced performance. This can be carried out by the inspection of wafer physical dimensions and its flatness, surface roughness, carbon and oxygen content, crystal defect and bulk resistivity (Tilli, 2010; Swaminathan, 2017; May, 2006). These methods are as follows.

4.4.1 Hot Point Probe

This method is used to determine whether a wafer is n-type or p-type. Two probes created the ohmic contact on the surface; one is kept heated (25°C–100°C) than other due to which the potential difference is created.

This difference is measured by using a voltmeter. The polarity difference in voltmeter tells about the type of wafer.

4.4.2 Four-Point Probe

This method is used to measure the resistivity of wafer. Using two probes can do this, but in this case the contact resistance associated with the probe and spread current around it is very important that is not simply accounted for by this method. Therefore, in this method, the outer two probes are used to measure the current flow and the inner two probes are used to measure the voltage drop by using a high impedance voltmeter. It eliminated the problem in voltage measurement with probe contacts because no current flows through these probes.

Another method to measure the resistance of wafers is the Hall effect measurement. This method is more powerful than the four-point probe method. By using this method, the material type, carrier concentration and carrier mobility can be measured separately.

4.4.3 Fourier Transform Infrared Spectroscopy

This method is used to identify the oxygen and carbon content in wafers. This content is developed in wafer fabricated using the CZ Crystal Growing method. It is obtained through the use of wavelengths of atoms. In FTIR, spectra is obtained by falling of IR light on wafer and it is shown that the oxygen peak at $1106 \, cm^{-1}$ and carbon peak at $607 \, cm^{-1}$ further, the silicon absorbs the other wavelength. If the concentration of oxygen is more in interstitial sites of silicon wafer, then it makes silicon dioxide inside the regions and further improves the yield strength up to 25% but when the concentration of carbon is more in wafer then it reduces the band gap of silicon. That may allow the formation of a new type of semiconductor in near future.

4.5 OXIDATION

It is a process to develop silicon dioxide on the surface of a silicon wafer. It can be achieved by heating of silicon wafer in an oxygen environment at high temperature because of the high band gap energy of silicon. It is one important process in the development of semiconductor devices. According to the thickness of the oxide layer, it is used in different applications such as if the thickness is 1 μm then it is used for field oxides, and if the thickness is 100 Å, then it is used as a controlling device for MOS. It is performed by two techniques; dry oxidation and wet oxidation. In dry oxidation, the silicon wafer is exposed to a pure dry atmosphere (1200°C) of oxygen. It produces a uniform layer of oxides, but it grows very slowly. In wet oxidation, the silicon wafer is exposed to a steam atmosphere (900°C–1000°C). It grows the oxide layer very fast and usually grows a thick oxide layer but due to the application of steam, hydrogen molecules are generated that develop the imperfection in the oxide layer and reduce its quality. The important parameters that need to be controlled are temperature, oxidizing species and cleanliness of the wafer (Tilli, 2010; Swaminathan, 2017).

4.6 DEPOSITION

It is a process to develop the thin film over the silicon substrate in the range of micrometer to few nanometers. The quality of deposition depends upon the composition of film, level of contamination, defect density (pinhole and other structural defects, it should be minimal), mechanical property (minimal stresses), optical property (reflectivity) and electrical property (conductivity). It is divided into two main categories, i.e., physical vapor deposition (PVD) and chemical vapor deposition (CVD) (Seshan and Schepis, 2018; Swaminathan, 2017).

4.6.1 Physical Vapor Deposition (PVD)

The basic principle of this deposition technique is melting and vaporization. The process is performed under vacuum. It involves the basic four steps, i.e., evaporation, transportation, reaction and deposition. It improves the hardness, wear and oxidation resistance of substrate due to which the life of the components is increased. It is used in almost every type of inorganic material and a few types of organic materials. High capital cost, high vacuum, high-skilled worker, high cooling arrangement and low rate of deposition are the limitations of this technique. It has a wide range of applications such as automobiles, aerospace, cutting tools and surgical tools. It is classified as follows.

4.6.1.1 Evaporative Deposition

In this technique, the targeted material is kept over the crucible and starts heating under high vapor pressure, due to which the material melts and starts vaporizing. The vaporized atoms fall over the substrate surface that is kept on the top and start the condensation, resulting in deposition on the substrate surface.

4.6.1.2 Electron Beam PVD

In this technique, the high-energy electron beam is used to melt the targeted material under high vacuum. The melted material is vaporized and deposited on the substrate surface.

4.6.1.3 Sputtering

This technique was developed by Langmuir in 1920 to deposit the thin film of metals such as Ni, Co, Au, Al and Ti on substrates. In this technique, the high energy of ions is bombarded on the targeted materials under high vacuum. After sufficient energetic particles on the surface, the atoms ejected and deposited on the substrate. It is carried out using a top-down approach (means targeted material is kept on top and substrate is kept on bottom). It can produce a uniform thickness. This technique is widely used in microelectronics for decorative and protective coatings, for pattern generation and for surface hardening.

4.6.2 Chemical Vapor Deposition (CVD)

In this technique, the reactant gases are introduced into the reaction chamber through forced convection that reacts with the heated substrate surface and forms a non-volatile solid film over the surface of substrates. The generated byproducts are further diffused into the main gas stream and transported through forced convection from the exhaust system. It consists of the system for gas delivery, reaction chamber, energy source, vacuum system, and process control equipment and exhaust system. This process can produce the deposition at a very high rate and also deposit materials that are very hard to evaporate. Further, it can grow the epitaxial thin film with good reproducibility. The corrosive and toxic gases, high temperature and complex phenomena are the main limitations of this process. It is widely used to deposit the thin film in semiconductors and related devices such as ICs, sensors and optoelectronic devices. It can also be used for the development of composite coating and powder production.

4.7 LITHOGRAPHY

It is a process of geometric pattern development on the wafer through a mask. The first step in this process is cleaning the wafer surface to remove the dust particles and any organic impurities. Then, the oxide layer is formed by either the dry or wet oxidation method. On a grown oxide layer, the UV-sensitive photoresist is coated followed by preheating. It is performed to remove the solvent from the surface. After this, the mask of metallic glass is placed and properly aligned on the surface of the wafer followed by high-intensity UV exposure for a few seconds for cross-linking of photoresist. Further, the developer is used on UV-exposed samples to remove the unexposed area of photoresist and develop the pattern on the wafer surface. The developed sample is kept for post-heating to improve the strength and adhesion of the developed pattern on the wafer surface. The two types of photoresist are used for the development of patterns; these are positive photoresist and negative photoresist. If the unexposed portion of photoresist is solidified and the exposed portion is washed away using the developer, then it is known as positive photoresist, while if the exposed portion of photoresist is solidified and the unexposed portion is washed away using the developer, then it is known as negative photoresist. The linewidth is the major issue in this area. It refers to the width of the smallest feature developed on the wafer. Currently, the minimum linewidth used is in the range of 0.12–0.25 μm. There are many forms of lithography available but the most commonly used type is photolithography (Crone, 2008; Groover, 2018; May, 2006).

4.7.1 Photolithography

It is a process that defines the shape of the micromachined structure on the silicon wafer. In this method, the first step is to develop the mask, i.e., typically chromium pattern on the glass plate. Then, the wafer is coated with the polymer photoresist using a spin coater. Afterward, the same process is used as discussed in lithography. Finally, by using the developer, the pattern is transferred from the mask to the photoresist layer by dissolving the unsolidified resist. This process is known as ashing. Mostly, the plasma ashing is used to remove the residue and non-volatile contaminants left after ashing using wet chemicals and acids. This process is repeated until the complete removal of residue and photoresist from the developed structure. Figure 4.1 shows the schematic representation of the photolithography process.

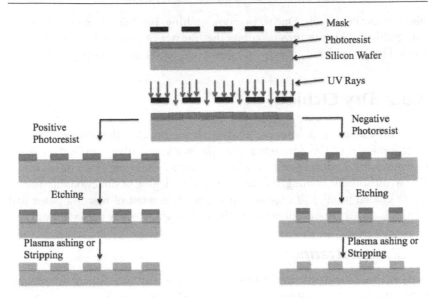

FIGURE 4.1 Schematic representation of the photolithography process.

4.8 ETCHING

This process plays an important role in the fabrication of microelectronic devices. It is a process where unwanted material is removed either by dissolving the structure in wet chemicals or by dissolving the structure with reactant gases in a plasma that form volatile products. Therefore, it is divided into two categories, i.e., wet etching and dry etching. The most important criteria of etching are selectivity. It is defined as the ability to etch one material without etching another with a controlled ratio of etch rates of two films (range 1:1 to 100:1). It typically lies in the range of 10–100 nm or more (Schwartz and Robbins, 1976; Swaminathan, 2017; Zhou, 2012; Groover, 2018).

4.8.1 Wet Etching

It is also known as isotropic etching. It means etching the materials at the same rate in all the directions. In this technique, the UV-exposed pattern is immersed in a liquid solution containing HF acid due to which silicon dioxide film etches very slowly. Due to isotropic etching, this leads to undercutting or side etching, which in turn prohibits the transfer of very high-resolution patterns. This process uses cheap equipment and produces a very high output.

4.8.2 Dry Etching

It is also known as anisotropic etching. It means etch the materials in one direction only. In this technique, the silicon dioxide film is etched by using energetic ions/reactant gases in a low-pressure system (i.e., vacuum system). Due to anisotropic etching, this leads to a high degree of directionality resulting in etched profiles. It requires only a small amount of reactant gases and provides a better linewidth control. There are various types of dry etching.

4.8.2.1 Sputtering

It is an anisotropic process. In this process, a stream of energetic ion beams is bombarded on the surface of the wafer under vacuum. These ions transfer their energy to surface atoms due to which atoms from the surface are removed due to excess of actual energy from binding energy. This process is known as sputtering. Ion current density and incident angle (i.e., ~ max. 80°) are very important parameters in this process, because if ion density is increased then the chance of irregular etching on the surface is more. Further, larger ion bombardment on the wafer surface causes poor profile and a reduction in selectivity to mask material.

4.8.2.2 Plasma Etching

It is an isotropic process. In this process, the etching is taken place by the reaction of silicon dioxide film with chlorine plasma or fluorine plasma generated by an RF (radio frequency) source. The process is similar to wet etching.

4.8.2.3 Reactive Ion Etching

It is a combination of sputtering and plasma etching process. Therefore, it is used for both the phenomenon of momentum transfer and chemical reaction to remove material from the wafer surface. The wafer is placed on a powered electrode and gives the negative bias to wafer due to which the accelerated positive charge ions move toward the wafer surface. These positively charged ions boost the chemical etching and allow anisotropic etching on the wafer surface.

4.8.2.4 Cryogenic Dry Etching

This process is used to etch very deep features with vertical walls. The workpiece is kept at cryogenic temperature and uses sulfur hexafluoride to provide fluorine ions for etching at this temperature. Therefore, silicon is removed in the form of silicon fluoride (SiF4), which is a volatile product. In this process, liquid nitrogen is used as a cooling agent for achieving temperature down to −110°C and helium gas is injected behind the wafer to provide good thermal contact. The temperature control in the whole feature of large wafers is difficult because of the variation in profile due to temperature variation. This problem can be overcome by providing multiple points for the injection of helium gas. It has developed a smoother feature but the rate of etching is very slow.

4.9 METALLIZATION

It is a process through which the components of ICs are interconnected by conducting materials, i.e., aluminum, or connect the distinct devices together through microscopic wires to develop a circuit. In this process, a thin film of metal is developed using either CVD or PVD technique. It acts like a conductor pattern for the interconnection of distinct elements on the chip. This process is widely used to interconnect thousands of bipolar devices, i.e., MOSFET (metal–oxide–semiconductor field-effect transistor), using a thin line metal pattern. Low resistivity, low contact resistance, easy to develop, easy to etch, mechanically stable, smoother surface and reliability for long-term operations are the basic requirements of this process. Further, metal (low resistivity) and polysilicon (medium resistivity) are the possible materials for metallization. As per the application, nitride and carbides are used for metallizing diffused barriers. However, aluminum, nitride, carbide and polysilicon are used for metallizing gates, interconnection and contacts. Among all materials, aluminum is widely used because of several advantages such as good resistance, conductivity and adherence, easily deposited, and good mechanical bond. But, it is suitable up to 500°C. After that, it is merged and penetrates the oxide layer to generate a short circuit. Another important point is to remember that if aluminum is reacted with gold that it makes a purple plague compound ($AuAl_2$) that creates the voids and weakens the bond on chips (Tilli, 2010; Bhushan, 2017; Kovacs et al. 1998).

4.10 WIRE BONDING

Wire bonding enables ICs to be connected electrically with the package such as flat panel display, keyboard, connector, antenna and sensor. It also provides the mechanical support to fragile ICs and environmental protection to ICs. It is a solid welding process where two metallic materials, i.e., thin wire and developed pad, through the metallization process are fetched into contact under heat and pressure or ultrasonic energy. It is divided into two groups, i.e., wedge bonding and capillary or ball bonding (Tilli, 2010; Bhushan, 2017).

4.10.1 Ball or Capillary Bonding

In this process, the capillary tool is loaded with gold and comes into contact with an electronic-flame-off system that generates the spark. It is melted gold at the tip of the tool then applies the pressure, ultrasonic energy and heat to accelerate the bond formation at the bond pad on chip side. Due to the surface tension, it forms a spherical- or ball-shaped bond then the tool is raised and repositioned on the substrate to form a second bond (known as stitch bond) through a thin wire known as loop. The shape of the secondary bond is either crescent or fishtail type. Finally, the wire is broken off, and the capillary tool is moved up. This step is repeated until to get the desired connection.

4.10.2 Wedge Bonding

In this process, the wedge tool, having an angle between 30°C and 60°C, is loaded with the aluminum wire and comes into contact with the chip pad and then applies the ultrasonic energy, heat and pressure to form a bond on chip. Now, the tool is raised and repositioned on the substrate to form a second bond through a loop formation. Finally, the wire is broken off using a wire clamp, and a wedge tool is moved up.

Advantages of Wire Bonding
1. It is highly flexible and easily programmed for interconnections.
2. It produces lower defects and higher yield interconnections.
3. Process is highly reliable and advanced in terms of tools, equipment and materials.
4. Process is very economical and produces stronger bonds.

Disadvantages of Wire Bonding
1. Interconnection rate is slower due to point-to-point processing.
2. It requires a large footprint for interconnections.
3. Longer length of interconnections degrades its electrical performance.

4.11 PACKAGING

It is a very essential part in the fabrication of semiconductors. It affects the basic functions of chips on a micro level and cost, power and performance on a macro level. It acts like a container to grasp the semiconductor die. Various packaging techniques are used in the semiconductor industry (Tilli, 2010; Bhushan, 2017).

4.11.1 Through Hole Mount Package

This package uses lead on the components, which are inserted into holes (known as plated through hole) presented on the surface of the printed circuit board. Then, soldering is performed from the opposite side. It produces a stronger mechanical bond. Currently, this process is kept for bulkier components that need surplus mounting strength such as electrolytic capacitors.

4.11.2 Surface Mount Package

In this package, the components are mounted directly on the surface of the printed circuit board. It is replaced Through Hole Mount package technology in large amounts for smaller components. This is possible due to the non-presence of leads or the smaller presence of leads. In this process, fewer holes are needed that are drilled using abrasive boards and components are placed on both sides of the printed circuit board. It provides lower resistance and inductance at interconnected places. These packages are more complex and cannot be used directly with the breadboards.

4.11.3 Chip-Scale Package

It is a type of IC chip carrier. It consists of a single die in which the package is directly mounted with an area of no more than 1.2 times the original die area. It comes in many package forms to meet the surface mountability and dimensional requirements such as wire bonded, flip-chip and leaded and ball grid array.

4.11.3.1 Wafer Level Chip-Scale Package

In this technique, the packaging is performed on the wafer followed by dicing. It allows direct connection without the use of wire to a printed circuit board. Then invert the die and perform soldering for connection. It is a widely used process to develop the interconnection package on the wafer of silicon. Further, a dielectric passivation polymer film is applied over the wafer surface to provide a mechanical stress relief for the attachment of balls and electrical isolation on the surface of the die.

4.11.3.2 Wire Bonded Ball Grid Array

In this process, the array of soldered balls is arranged in the form of a grid. Then, the external pad frame with wire of bonded gold is stitched to the bond pad from the top-level padring layout. Further, these are connected to the ball using conductor traces on an interpose substrate.

4.11.3.3 Flip-Chip Ball Grid Array

In this process, the chip/die is placed in the flipped form, due to which the external ball grid arrays are connected to bond pads in the padring layout through conductor traces. At the time of packaging, on bond pads (die), the solder bumps are developed that are aligned and contacted with conductor traces when they are flipped. The bumped die is kept on the substrate of the package where it is connected to the package using either pins or balls. This technique is used to connect the die bond pad to packages without using wire bonds. This method improves the current distribution by minimizing the voltage drop. Further, it is supported by the higher frequency design due to the lower inductance.

4.12 MEMS TECHNOLOGY

The MEMS is the acronym of Micro Electro Mechanical Systems. It is the most promising technology in the 21st century that has the potential to modernize the industry and consumer products. This technology is the combination of mechanical functions such as sensing, heating and moving and electrical functions such as switching on the wafer chip fabricated through micro fabrication techniques. It has played an important role in society due to the development of ICs. These are used as a base for calculators, PCs, information systems, telecommunications, robotics, space travel, defense weapons, etc. Therefore, the fabrication of microelectronic devices/components like transistors, diodes, resistors and capacitors results in placing of more components onto an IC and this increases the overall efficiency of the system (Tay, 2012; Swaminathan, 2017; Madou, 2002; Groover, 2018).

4.12.1 Materials for MEMS

There are various materials used to develop the MEMS device. These are silicon, gallium arsenide (GaAs), quartz and piezoelectric crystals. Among all materials, silicon is an ideal material for MEMS device fabrication, but its brittleness property and hydrophilic surface behavior restricts its usage in many applications and requires more sophisticated handling. Due to the aforementioned limitations, researchers have tried to develop alternative materials for ease of MEMS fabrication. They found some polymers such as PMMA (poly-methyl methacrylate) and SU-8, which are most suitable for the development of MEMS. Among all, SU-8 showed a promising potential to be used in the development of MEMS. SU-8 is a thermoset polymer that consists of three components: bisphenol A novolavglycidyl epoxy resin, gamma-butyrolactone and 10 wt.% of mixed triarylsulfonium known as photo-acidic generator. It is an epoxy-based negative photoresist that is considered a material for the next-generation fabrication of 3D MEMS structures such as micro-engines, accelerometers and micro-gears. These 3D MEMS structures are potentially used in automotive, bio-medical, semiconductor and aerospace industries. SU-8 is very superior to Si in terms of properties such as UV curability, biocompatibility, higher thermal capability (~300°C), easy fabrication for high-aspect-ratio parts, good optical properties, cost-effectiveness and lower surface energy, but it has few disadvantages such as poor mechanical properties and poor tribological properties that have restricted its usage in MEMS applications (Jiguet et al., 2006; Katiyar et al., 2016; Rathaur et al. 2019; Katiyar et al., 2020).

4.12.2 Fabrication Process

Micromachining is used as a fabrication technique to shape silicon for the realization of 3D miniature components and mechanical devices that are well suited to microelectronic devices. The micromachining techniques are bulk micromachining, surface micromachining and high-aspect-ratio microma-chining, i.e., LIGA (Lithographie, Galvanoformung, Abformung, meaning lithography, electroforming and molding). Micromachining has several advantages such as lower manufacturing cost and lighter parts makes them very useful in sensors and actuators. But it has few limitations such as complex design and packaging and not fully developed methods (Kim et al, 2007; Kovacs et al., 1998; Madou, 2002; Tai, 2012).

4.12.2.1 Bulk Micromachining

This technique involves the bulk amount of removal from the surface. It is a subtractive process that is achieved by either wet etching or dry etching techniques. It is used to fabricate the channels, groves, pits, etc. The wet etching process is preferred for silicon and quartz to remove the unwanted material, but dry etching is preferred for silicon, metals, ceramics and plastic to remove the unwanted material.

4.12.2.2 Surface Micromachining

This technique was introduced in 1980, and it is performed on the surface to develop the microstructures. In this process, the thin sacrificial layer is developed using a pattern and photolithography method over the silicon wafer. Then, the structural material is deposited using thin film deposition techniques. Further, these developed layers are removed through an etching process to get a desired freestanding microstructure. Therefore, this process involves two distinct types of materials. One is structural material (polysilicon, silicon nitride and aluminum) and another material (oxide) for a sacrificial layer of freestanding mechanical structure. The success of surface micromachining largely depends upon the efficient removal of the sacrificial layer due to which structure can be free stand for actuation purposes.

4.12.2.3 LIGA

It is an important technique to develop the high-aspect-ratio microstructures. In this process, an acrylic sheet of PMMA with the desired shape of mask is kept under the X-ray radiation due to which the exposed area is weakened and dissolves chemically. In the left our pattern on PMMA, the metal is electroformed thereafter the defined tool insert for subsequent molding steps. This process is capable of developing the height of microstructures up to 1000 µm. The requirement of X-ray in forming microstructures is the only problem that restricts its wide applications. To overcome this issue, the sacrificial LIGA is developed where X-ray is not used. In this process, thick PMMA is replaced with polyimide as electroplating mold. It opens the production of MEMS devices with the lowest manufacturing infrastructures that means investment and facilities (Strohrmann et al., 1994).

4.12.3 Application of MEMS

In the 21st century, the MEMS has played a very important role in the large volumes of applications in distinct areas such as defense, automobile, medical, electronic and communications. Few examples of microengine and microactuator are shown in Figure 4.2 that are fabricated on silicon wafer and SU-8 nanocomposite using photolithography followed by surface micromachining technique as discussed above (Swaminathan, 2017; Groover, 2018; Katiyar et al., 2020; Strohrmann et al., 1994).

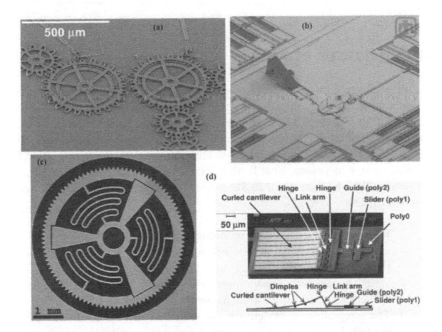

FIGURE 4.2 (a) Gear meshing and (b) drive and mirror systems, produced on Si wafer using surface micromachining in Sandia National Laboratories SUMMiT™ process (reproduced with permission from Kim et al., 2007). (c) Gear wheel of 5 mm diameter, developed using SU-8 nanocomposites (reproduced with permission from Jiguet et al., 2006). (d) Electrostatic lateral output motor (scanning electron micrograph and schematic of cross-sectional) with cantilever dimensions of 350 mm × 196 mm (reproduced with permission from Kim et al., 2007).

Medical: MEMS devices are used as blood pressure sensor, muscle stimulators and drug delivery systems, implanted pressure sensors, pacemakers, prosthetics and miniaturize diagnostic instruments,

Defense: MEMS devices are used as arming systems, surveillance, data storage, aircraft control and weaponries guidance.

Automobile: MEMS devices are used as sensors for internal navigation, sensors for brake force, sensors for air-conditioning compressor, sensors for fuel level and vapor pressure, intelligent tires and suspension control accelerometer.

Electronics: MEMS devices are used as inkjet printer head and disk drive head, sensors for earthquake, sensors for avionic pressure, projection screen in televisions and mass data storage systems.

Communications: MEMS devices are used as voltage control oscillators, RF relays, switches and filters, fiber-optic network components, tunable lasers, projection displays in portable communication devices and instrumentation, splitters and couplers.

4.12.4 Issues of MEMS

As it is well known that this technology is manufactured by tiny ICs using ICs fabrication techniques and it is not about miniaturization. Therefore, in order to develop the successful MEMS device, the basic principles of physics with laws of scaling require being fully understood at both scales, i.e., macro and micro. This is necessary because by decreasing the size, the surface area to volume ratio is increased. That further increased the adhesion on the surface (Bhushan, 2017; Katiyar et al., 2020). There are also some other issues that are listed out here.

1. In these devices, friction is more than inertia due to the higher adhesion on the surface, due to which more wear on the surface causes the reduction of device life.
2. These devices have shown more heat dissipation than heat storage. Sometimes, it is useful or not as per the application.
3. The integration of the chip is complex and specific to the device domain. Further, the cost of MEMS devices is another issue that can be overcome by producing the devices in mass production.
4. Testing facilities are not well established for such types of miniaturized devices. These are more costly and complex with the comparison of conventional ICs.

4.12.5 Future of MEMS

MEMS is a very emerging technology that makes revolutionary changes in both the industry and human life through the development of more sensitive structures for sensing, imaging, etc. The miniaturization of components has added greater benefits to our society by reducing the component cost, easily transported, improved efficiency and highly precise surgical instruments (Bhushan 2017; Tai, 2012). But it can be only achieved by addressing the below points effectively:

- To develop the fabrication facilities and infrastructure for MEMS devices.
- To develop the design, simulation and modeling software due to which the trial-and-error method for MEMS fabrication can be eliminated, and before the development phase itself the problem is rectified.
- Packaging and testing are another area in MEMS fabrication that adds more fabrication cost. More sophisticated tools and facilities are required. Further, standardization is also required to develop MEMS devices in large amounts.

4.13 NEMS TECHNOLOGY

The NEMS is the acronym of Nano Electro Mechanical Systems. It is another most promising technology in the 21st century that has the potential to modernize the industry and consumer products. This technology is also the combination of mechanical functions such as sensing, heating and moving and electrical functions such as switching on the wafer chip fabricated at the nano scale ($\leq 100\,nm$) (Drexler, 1992). It is typically integrated with nano-electronics with mechanical actuators. The NEMS devices have very low mass and high mechanical frequencies due to which it is a very promising technology in the analyses of small displacement and extremely weak forces at the molecular level (Roukes, 2000; Craighead, 2000). Further, this technology reduces the power consumption cost and production cost that further offers the enormous prospective for new applications and primary measurements. The fabrication method of NEMS technology is based on two approaches. Top-down approach is utilized the submicron fabrication technique (i.e., lithography) to develop the structure from the bulk material. It may be thin film or thick film. Another approach is bottom-up, which utilizes the deposition phenomenon to develop the nanoscale devices (Ekinci and Roukes, 2005; Kim et al., 2007; Zhou, 2012).

4.13.1 Materials and Fabrication of NEMS

The materials used in NEMS technology are silicon, silicon nitride, silicon carbide and gallium arsenide for development of ICs. A few devices are developed using a top-down approach that is a nanomechanical beam and paddle oscillator. Further, advancement is carried out by the inclusion of nanowires or nanotubes in NEMS devices through the bottom-up approach. For example, Yang et al. (2001) fabricated the doubly clamped silicon carbide beam for mechanical actuation of resonant frequency from 2 to 134 MHz using the top-down approach (Figure 4.3).

4.13.2 Application of NEMS

NEMS technology is still in an initial stage of development. The potential applications of NEMS devices are Bio-NEMS for the detection of chemical forces using atomic force microscopy (Bhushan, 2017), mechanically detected magnetic resonance imaging (Drexler, 1992), neurological surgical instruments and pressure sensor for blood pressure measurement (Tan et al., 2017), accelerometer for measurement of tilt and orientation in mobile phones,

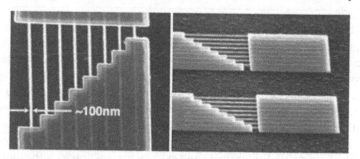

FIGURE 4.3   Fabricated doubly clamped silicon carbide (SiC) beam for actuation of resonant frequencies from 2 to 134 MHz using the top-down approach. It was patterned at Caltech from 3C-SiC epilayers grown at the Case Western Reserve University (reproduced with permission from Yang et al., 2001).

nano-nozzle for ink direction for inkjet printer and thermal actuators for the measurement of small thermal expansion.

4.13.3 Future of NEMS

NEMS is a very emerging technology that makes revolutionary changes in both the industry and human life through the development of more sensitive structures for sensing, imaging, etc., at the nano level. This development creates new facilities and very sophisticated equipment for experimentation at the nano scale. Nowadays, testing of such devices is performed using atomic force microscopy. Further, the development of nanotubes and nanowires has created new possibilities in the development of NEMS devices because of their enhanced mechanical properties (Bhushan, 2017). NEMS systems can be integrated with the electronics and optics for the fabrication of highly functional devices such as nano robot and nano motor create new possibilities (Orhan et al., 2020). Moreover, biochemistry is another area where NEMS technology can play a very important role in analyzing the surface chemistry of molecular systems (Crone, 2008; Kim et al., 2007; Tan et al. 2017). Therefore, NEMS technology will be surely anticipated in a decade or so through the development of highly sensitive and functional devices with a greater mechanical strength.

4.14 SUMMARY

It is found that silicon is the most widely used material in the fabrication of semiconductor devices but its hydrophilic nature and brittleness property restricts usage in many complex shape devices. To overcome this issue, IBM in 1980 developed the polymer materials SU-8. It is a thermoset polymer with more superior properties than silicon. It is showing a growing interest and more promising materials for the fabrication of MEMS/NEMS devices. These devices have the potential to sense mass, displacement, impact, etc. Such devices utilize the concepts of design, engineering and manufacturing expertise from a wide spectrum of technical areas such as the development of IC, material science, mechanical engineering, chemical engineering, electrical and electronic engineering, as well as optics, fluid and packaging.

REFERENCES

Bhushan, B. (2017) *MEMS/NEMS and BioMEMS/BioNEMS: Tribology, Mechanics, Materials and Devices.* Springer Handbooks, Berlin, Heidelberg, 1331–1416.

Craighead H. G. (2000) Nano electro mechanical systems, issues in nanotechnology, Science 290, 1532–1535.

Crone W. C. (2008) A Brief Introduction to MEMS and NEMS, William N. Sharpe Jr., Eds., *Springer Handbook of Experimental Solid Mechanics*, 203–228.

Drexler K.E. (1992) *Nanosystems: Molecular Machinery, Manufacturing and Computation*, Wiley, New York.

Ekinci K. L. and Roukes M. L. (2005) Nanoelectromechanical systems, *Review of Scientific Instruments*, 76, 061101.

Groover M. P., (2018) Principles of Modern Manufacturing SI Version, Wiley India.

Jiguet S., Judelewicz M., Mischler S., Bertch A., Renaud P. (2006) Effect of filler behavior on nanocomposite SU8 photoresist for moving micro-parts, *Microelectronic Engineering*, 83, 1273–1276.

Katiyar J. K, Sinha S. K., Hirayama T. and Kumar A. (2020) Tribological analysis of tip-cantilever made of SU-8/talc/PFPE composite, tribology, *Materials, Surfaces & Interfaces*, 14(2), 92–101.

Katiyar J. K, Sinha S. K. and Kumar A., (2016) Friction and wear durability study of SU-8 composites with talc and graphite as fillers, *Wear*, 362–363, 199–208.

Kim S. H., Asay D. B., and Dugger M. T., (2007) Nanotribology and MEMS, *Nano Today*, 2(5), 22–29.

Kovacs G. T. A., N. I. Maluf, and K. E. Petersen, (1998) Bulk micromachining of silicon, *Proceeding of the IEEE*, 86(8), 1536–1551

Madou M. J. (2002) *Fundamentals of micro fabrication: The Science of Miniaturization*, Second Edition, CRC Press, USA.

May G. S., and Spanos C. J., (2006) *Fundamentals of Semiconductor Manufacturing and Process Control*, Wiley-IEEE Press; 1st edition.

Orhan E, Yuksel M, Ari AB, Yanik C, Hatipoglu U, Yagci AM and Hanay MS (2020) Performance of nano- electromechanical systems as nanoparticle position sensors. *Frontiers in Mechanical Engineering*, 6, 37.

Petersen K. E., (1982), Silicon as a mechanical material, *Proceeding of the IEEE*, 5: 420–457.

Rathaur A. S., Katiyar J. K. and Patel V. K., (2019) Thermo-mechanical and tribological properties of SU-8/h-BN Composite with SN150/Perflouropolyether as filler, *Friction*, 8(1), 151–163.

Roukes, M. L. (2000) Nano Electro Mechanical Systems, Technical Digest of the 2000 Solid-State Sensor and Actuator Workshop, Hilton Head Isl., SC, 6/4–8/2000 (ISBN 0–9640024-3-4)

Schwartz, B. and Robbins, H. (1976), Chemical etching of silicon, VI. Etching technology, *J. Electrochem. Soc.*, 123, 1903.

Seshan K. and Schepis D. (2018) *Handbook of Thin Film Deposition*, Elsevier Inc.

Strohrmann M., P. Bley, O. Fromhein, and J. Mohr, (1994), Acceleration sensor with integrated compensation of temperature effects fabricated by the LIGA process, *Sensors and Actuators, A* 42(1–3): 426–429

Swaminathan P., (2017) *Semiconductor Materials, Devices and Fabrication*, Wiley India.

Tai Y. C. (2012) Introduction to MEMS, *Microsystems and Nanotechnology*, 187–206.

Tan A., Jeyaraj R. and De Lacey S.F. (2017) Nanotechnology in neurosurgical oncology, *Nanotechnology in Cancer*, 139–170, Elsevier.

Tilli M. (2010) Handbook of silicon based mems materials and technologies, *Micro and Nano Technologies*, Elsevier, 71–88.

Yang Y. T., Ekinci K. L., Huang X. M. H., Schiavone L. M., and Roukes M. L. (2001) Monocrystalline silicon carbide nanoelectromechanical systems, *Applied Physic Letters*, 78, 162.

Zhou Z., Wang Z., Lin L. (2012) *Microsystems and Nanotechnology*. Springer, Berlin, Heidelberg, 187–206.

A New Vista of Manufacturing Technology for Industry 4.0

5

5.1 BACKGROUND AND IMPACT OF INDUSTRIAL REVOLUTIONS

The three industrial revolutions have been observed since the 18th century, and these revolutions were expressed as innovative jumps in industrial processes that could lead to higher productivity. The first industrial revolution is witnessed as the period where the mechanical production assets based on water and steam power were introduced. This would lead to the industrial transformation of society with trains and mechanization of manufacturing. The efficiency of the first industrial revolution was also improved via the use of water and steam power as well as the development of machine tools. The second industrial revolution is usually seen as the period where electricity and new manufacturing inventions such as the assembly line led to the area of mass production and to some extent to automation. The third industrial revolution is observed as the period where electronics and information technology for higher automation of production was introduced. The fourth industrial revolution is started in the 21st century, wherein the cyber-physical systems (CPS) in manufacturing have introduced. The CPS are those physical systems/objects, which are being seamlessly integrated with the information network. In the fourth industrial revolution, people are experiencing a digital transformation where changes connected with innovation in the field of digital technology in all aspects of society and economy. This digital trend is influencing the way products are manufactured and services are offered. In this revolution, the internet is integrating with smart machines, manufacturing systems and processes to create a sophisticated network. The evolution of industrial revolutions from first to fourth is shown in Figure 5.1.

The impact of each industrial revolution includes

   i. Introduction of new products and means of producing the existing ones.
  ii. Disruption of the competitive existing state of affairs (both within and between countries and enterprises).
 iii. New requirements to people engaged in or available for work in industry and infrastructure.

5.2 INTRODUCTION TO INDUSTRY 4.0

Industry 4.0 is derived from the German term "Industrie 4.0". It was first used in a project in the high technology planning to change German manufacturing in which the Internet of things (IoTs) and CPS took center stage, along with a further focus on production, people, environment and security. Industry 4.0 (I4.0) is explicitly connected with the advent of a fourth industrial revolution, driven by the internet, IoTs, big data, cloud, robotics,

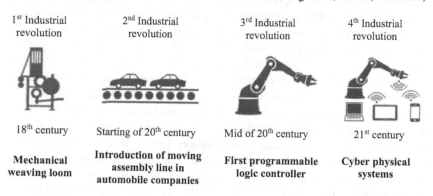

1st Industrial revolution	2nd Industrial revolution	3rd Industrial revolution	4th Industrial revolution
18th century	Starting of 20th century	Mid of 20th century	21st century
Mechanical weaving loom	Introduction of moving assembly line in automobile companies	First programmable logic controller	Cyber physical systems

FIGURE 5.1 Evolution of the first to fourth industrial revolutions.

artificial intelligence, etc. Industry 4.0 highlights the idea of constant digitization and connecting all productive units in an economy.

Different people have different opinions about Industry 4.0. Some people say that Industry 4.0 is also known as the smart industry that refers to the technological evolution from embedded systems to CPS. Some other people say that I4.0 represents the coming fourth industrial revolution on the way to IoTs, data and services. But, in general, Industry 4.0 refers to the transformation of the manufacturing industry through the intelligent networking of machines and processes for industry with the help of information and communication technology. The term is used interchangeably with the fourth industrial revolution in industry. Industry 4.0 bridges the gap between the physical and digital world through CPS, enabled by IoTs.

Industry 4.0 is the evolution to CPS (i.e., CPS form the basis of Industry 4.0) showing the fourth industrial revolution on the road to an end-to-end value chain with Industrial IoTs and decentralized intelligence in manufacturing. Decentralized intelligence helps to create intelligent object networking and independent process management, with the interaction of the real and virtual worlds representing a crucial new aspect of the manufacturing and production process. In other words, the CPS use advanced industrial control systems, embedded software systems and arrange an internet address to connect and be addressed via IoTs. This way, products and means of manufacturing get networked and can communicate, enabling new ways of production, value creation and real-time optimization.

The design principles of Industry 4.0 enable identifying and realizing applications in industrial transformation. The principles are interoperability, virtualization, decentralization, real-time capability, service orientation and modularity. The characteristics of I4.0 include a high degree of individualization of products with highly flexible production, early involvement of customers and business partners in design and value creation processes and connecting production and high-quality services would result in hybrid products.

5.3 TRANSFORMING LANDSCAPE IN MODERN MANUFACTURING

Over the decades, manufacturing engineering existing as the backbone of modern technology-driven growth and development of industries and has remained at all times in the leading position of the engineering and technology revolution. However, to date, for a country to become top nation in the industrial world, the manufacturing sectors (under the umbrella of the industrial sector) contribution has to increase to still higher levels in order to enhance the country's GDP. The product customization and personalization will become more and more demanding necessitating quicker responses, where total mastery over the new tools of manufacturing will have to be ensured, rather than just familiarity and learn as you go attitude. The advancements in artificial intelligence, cloud computing, machine learning and automation are impacting manufacturing factory operations like never before. Readiness to face such and similar challenges is crucial. Thus, modern manufacturing technologies have progressed to push the expedited growth and development of a new industrial revolution, i.e., Industry 4.0.

Industry 4.0 in the manufacturing sector is considered a new paradigm, which is a collection of cutting-edge technologies that support effective and accurate engineering decision-making in real time that delivers results. Industry 4.0 can change the landscape in manufacturing. Industry 4.0 can use technology to move from mass production to mass customized production, and it currently going at a rapid pace. This allows the manufacturing industries to react to customer demand more efficiently and directly. The key elements that are driving in today's manufacturing world are shown in Table 5.1.

In recent years, the manufacturing technology has greatly contributed to the manufacturing of intricate components and systems, miniature electronic circuits and devices, ultrafine features and materials, etc. that encouraged the accelerated growth of advanced products making it possible for the humankind to meet their current stringent requirements. The potentiality of innovation and strong scientific explorations has led to the new vista of digital manufacturing technologies. The industries with huge automation and IoTs of everything would witness a sea change in their workforce that can be comprised of workers, robots and intelligent machines working together as a winning team of

TABLE 5.1 Key elements driving in today's manufacturing

NEW MATERIALS	NEW DESIGN METHODS	PRODUCT INNOVATIONS
Intelligent Materials	Life cycle assessment design	Smart products
Ion exchange Materials	Design for manufacturing, environment, recycle and reuse	Multi-functionality
Nanomaterials	Integrated design and manufacturing	Recycle and reuse
Biomaterials		Ever-reducing product life cycle

the knowledge workers. As a result, a new kind of quality and reliability at all levels of manufacturing will have to be identified in line with the fast and connected IoTs in the industrial world. The decision-making in such a networked environment will put a demand on the use of various digital techniques.

Industry 4.0 encompasses various digital technologies, automation and data exchange techniques, which revolutionize the industrial modern manufacturing technologies. This includes CPS, additive manufacturing (3D printing), nanotechnology/advanced materials, ultrafine electronic devices, connected and networked with global systems of advanced robotics, automation, artificial intelligence, machine learning, IoTs, cloud computing, automation augmented reality, virtual reality, simulation, cyber security and big data and analytics that can change the waves of unprecedented development (Boston Consulting Group Inc., 2016). These techniques monitor physical processes, create a virtual copy of the physical world and make decentralized decisions. Over the IoTs, CPS communicate and coordinate with each other and with humans in real time, and via the internet of services, both internal and cross organizational services are offered and used by manufacturers of the value chain. Using real-time data to monitor machines will lead to overall equipment effectiveness, culminating in better quality standards, shorter cycle times and more productivity. This will result in a sudden increase in new creativity by humans.

The I4.0 trend is witnessing as a transforming force that will greatly impact the industry. The trend is building on digital technologies, which will lead to connected farms, optimization of resources, enhanced quality and quantity of yield, environmental protection, increased income, better standard of living and sustainable rural development. The manufacturing excellence of Industry 4.0 is shown in Figure 5.2. The figure shows an overall view of a company as interconnected global systems of digital technologies, automation and data exchange techniques. Outside of the company there found to be supplier network, future resources, new customer demands and the means to meet them, whereas inside of the company there seem to be modern and new manufacturing technologies, new materials and new ways of storing, processing and sharing data (Roland Berger strategy consultants, 2014).

The above developments will prompt the industrial researchers to go on to molecular manufacturing in perspective. As a result, the waste and the energy consumption in manufacturing would be minimized. This could be achieved – when the life cycle of the products ends they start to disintegrate. The disintegration reverts back to the molecules of materials that were accumulated via molecular manufacturing into different forms and could be performed as engineered products. This will lead to no waste remains to dispose off. Similarly, the new energy systems will grow on the ever-expanding utilization of renewable energy sources, e.g., solar energy.

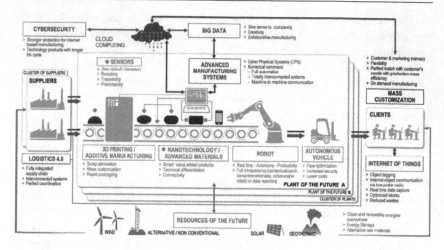

FIGURE 5.2 Manufacturing excellence of Industry 4.0 (from Wilfried A. and Harsh V.S., 2014).

Industry 4.0 can offer various benefits to modern manufacturing processes such as

i. Increased productivity through optimization and automation;
ii. Increased flexibility;
iii. Increased quality and speed;
iv. Real-time data for a real-time supply chain in a real-time economy;
v. Real-time monitoring and IoTs-enabled quality improvement;
vi. Better quality products;
vii. Better working conditions and sustainability.

Industry 4.0 is an opportunity to change the economic rules of the industry, especially to overcome the dependence of the country on the labor arbitrage-based manufacturing work. Industry 4.0 will change the game for industrial users, infrastructure suppliers and technology providers. The various approaches to take up the challenge of Industry 4.0 include

i. Industry 4.0 covers a broad set of technologies with a huge field for innovation and creative solutions. The renowned industry business models will create new opportunities for adding value, but those will depend on breakthrough innovations for technology and the ability to bring them to market.
ii. Develop future champions, i.e., to develop future technology planning and bridge the gaps.

iii. Establish a dynamic infrastructure for reliable power supply, telecommunication supply, and large data transfer and security procedures.

iv. Foster new talent.

Nowadays, the interdisciplinary behavior of manufacturing engineering can play a major role in the industry and the manufacturing engineers should take the ideas and understand them fully so that the technological innovations can be modulated into new products and systems. This will lead to the creation of new markets and overcome the stringent problems of society around the world. Manufacturing engineering will remain the core of engineering activity in all times to come and will continue to have impact on the human civilization to its future glory.

However, Industry 4.0 will make an impact on all job levels in the industry which not only helps to land in a desired job but it supports to grow as an individual, which results in growth in the country's economy. The list of factors that I4.0 may force an individual to remain employable include – update on the latest technology, self-evaluation, soft skills and adaptability. The top skills relevant to I4.0 include complex problem solving, critical thinking, creativity, people management, coordinating with others, emotional intelligence, judgment and decision-making, service orientation, negotiation and cognitive flexibility. In addition, the digital manufacturing will change the middle-level engineering manpower but will add technical graduates and PhD-level employees to the workforce.

5.4 MAJOR CHALLENGES

The current practical challenge for manufacturing engineers is to design and develop solar cells of high efficiency of about 95% against 18% at present. The solution to the problems of seawater desalination using graphene nanostructured material and producing steam at room temperature from seawater is a challenge to the industry. This will provide a hope for the security of water to billions of mobs around the globe. The explosive power of high detonation capacity nanoparticles would fuel the manufacturing engineer's prediction to design and develop inexpensive innovated systems for industrial advancement to drive I4.0. The research and innovations in new material technology interconnected with advanced manufacturing systems automation along with various digital technologies would excite the imagination of young inspired minds for accelerating manufacturing technology advancement. The future of the industrial society would be essentially connected to our ability to innovate products, new materials and new design and manufacturing. The technological innovations and advancements will open up new pleasing views of manufacturing engineering in the future. These shall create the great happiness of I4.0. By then, the manufacturing engineering will cover a large canvas of society utility engineering.

5.5 CASE STUDY

There are a number of case studies are available, which are presented by various companies related to the utilization and realization of the digital transformation and leveraging into the competitive company. For example, an apparel company developed a model, wherein a customer goes to their various showrooms, gives his measurements and preference for the shirt and the same is delivered to the customer at the company price within the assured time. This could avoid the selling of just only standard sizes available in their stores. This customized manufacturing system could be possible only due to the digital platform, i.e., virtually model any design and develop the established process for effective customized production.

5.6 SUMMARY

In this chapter, the background and impact of industrial revolutions in a nutshell followed by the preamble to Industry 4.0 were briefly discussed. The changing landscape in modern manufacturing due to the advent of Industry 4.0 was briefly discussed. The challenges of the integration of automation, additive manufacturing and smart materials to meet the paradigm shift issues in digital manufacturing were highlighted.

REFERENCES

Industry 4.0: fourth industrial revolution guide to Industrie 4.0, www.i-scoop.eu/industry-4-0/.
Wilfried A. and Harsh V.S. Next Gen Manufacturing: Industry 4.0, Confederation of Indian Industry, Roland Berger Strategy consultants, 2014.

Index

advanced finishing technique 42, 43
advanced machining technique 38, 39
aerosol synthesis 80
application of MEMS 111
application of NEMS 114
applications of metal 3D printing 64

background of additive manufacturing 47
ball milling 75
blowing 21
boiling flask-3-neck 81
bound powder extrusion 61
bulk micromachining 110

capillary bonding 106
casting techniques 3
centrifugal casting 21
ceramic forming 19
ceramic mold casting 6
ceramic shell investment casting 4
ceramic-based nanocomposites 35
chemical vapor
    condensation 80
    deposition 101
    infiltration 35
chip-scale package 107
classification of MMTs 2
classification of nanomaterials 70
combustion flame 78
composites manufacturing techniques 29
contact Method 15
cryogenic dry etching 105
CZ Crystal Growing 98

deposition 34, 100
diffusion
    bonding 34
    technique 96
direct energy deposition technique 58
direct melt oxidation 35
drawing process 20
dry etching 104

electrochemical technique 83
electro-hydraulic forming 16
electrolyte deposition 34
electromagnetic forming 16
electron beam
    assisted machining 40
    PVD 101
    welding 27
etching 103
evaporative deposition 101
expandable pattern casting 5
explosive forming 15
explosive welding 26

finishing techniques 41
flip-chip ball grid array 108
float method 20
floating zone 98
forming techniques 14
Fourier Transform Infrared Spectroscopy 99
four-point probe 99
friction stir welding 25
fused deposition modeling 52

glass forming 20

hot point probe 98

incremental sheet metal forming 21
industrial revolutions 119
Industry 4.0, 120
inert gas condensation 77
infiltration 30
ion beam machining 40
ion implantation 96

joining techniques 22

laminated object manufacturing 51
laser beam welding 27
laser-assisted machining 40
LIGA 110

liquid silicon infiltration 35
lithography 102

machining techniques 36
magnetic pulse welding 24
materials 97
   for MEMS 109
MEMS technology 108
metal additive manufacturing techniques 55
metal binder jetting technique 60
metal-based nanocomposites 30
metallization 105
micro-ECDM 86
micro-EDM 84
micro-emulsion 79
microfluidic reactor 81
microwave welding 28
multi-jet modeling 53

nanoparticles 71
   applications 74
   production techniques 44, 75
need for MMTs 1
non-metal additive manufacturing
         techniques 48
nanotechnology 69
   concept in advanced machining 81
NEMS technology 113

oxidation 99

packaging 107
peen forming 17
photolithography 102
physical vapor deposition 100
plasma etching 105
plasma processing 78
plasma-assisted machining 37
plaster mold casting 5
polymer infiltration and pyrolysis 35
polymer-based nanocomposites 30
polyol technique 80
powder bed fusion technique 57
pressing 21
pulsed laser ablation 82

rapid solidification 9
reactive ion etching 104
rheo casting 13
rolling process 20

sagging 21
selective laser sintering 50
semiconductor 95
   doping 96
semisolid casting 11
silicon wafer fabrication 97
single crystal casting 10
sintering process 34
slurry infiltration 36
slush casting 7
sol-gel
   infiltration 36
   processing 79
solid ground curing 52
solution precipitation 79
spin casting 9
spray deposition 34
spray pyrolysis 78
sputtering 77, 101, 104
squeeze casting 8
standoff method 15
stereolithography 50
stir casting 11
strain-induced melt activation 14
submerged arc nanoparticle synthesis
         system 78
superplastic forming 18
surface micromachining 110
surface mount package 107

thermal-assisted techniques 37
thixo casting 12
thixo forming 19
thixo molding 14
3D ink-jet printing 53
Through Hole Mount package 107
transforming landscape in modern
         manufacturing 121

vacuum mold casting 7
vapor deposition 35
various welding processes 23

wafer level chip-scale package 108
wedge bonding 106
wet etching 103
wire arc additive manufacturing technique 62
wire bonded ball grid array 108
wire bonding 106
wire explosion 77